CONTENTS

INTRODUCTION

Trees and shrubs are such a feature of the British scene that many people take them for granted. However, imagine what the landscape (not to mention the parks and gardens) of Great Britain and Ireland would be like without them and you can begin to appreciate their significance. Indeed, in many ways it is our trees and shrubs, more than other living things, that help define what we think of as the British countryside.

Native tree species have an ecological significance that goes beyond their individual presence. The role they play in the ecology of our native woodlands particularly fascinates me, perhaps more than their appearance. So I make no apologies for this book being biased in favour of native species. However, introduced trees and shrubs also have a role to play in today's world. Some are widely naturalised, many soften our otherwise often brutal urban landscapes and still more are familiar and valued features of mature gardens. Consequently, I have included a wide range of familiar planted species, along with a selection of more unusual or exotic trees and shrubs mostly associated with collections and arboreta.

ABOVE: Whether planted or growing in the wild, the Common Beech produces autumn colours that cannot fail to lift the spirits.

RIGHT: Known best for its colourful berries, the Rowan is widely planted in urban settings and is a welcome sight in autumn.

My personal interest in trees and shrubs extends beyond enjoyment of trees for their own sake and their role in our ecology. I am also fascinated by the uses of timber and tree products in woodland crafts and traditional practices. Sections of the book reflect this interest.

THE REGION COVERED BY THIS BOOK
The region covered by the book comprises the whole of mainland England, Wales, Scotland and Ireland, as well as offshore islands including the Shetlands, the Orkneys, the Hebrides, the Isle of Man, the Isles of Scilly and the Channel Islands.

THE CHOICE OF SPECIES
The coverage of the book is restricted mainly to what most people understand to be trees and larger shrubs; the former are usually defined as single-boled plants with a trunk that exceeds 5m, while shrubs are typically multi-stemmed. However, for the sake of completeness, and as a minor self-indulgence, I have also included native members of tree groups such as willows and birches that should not qualify for inclusion, in the strict sense, on the grounds that they are too small.

Complete British Trees will enable amateur naturalists to identify all native and widely naturalised tree and shrub species found growing wild in the British countryside. With an eye to the exotic, it also allows naturalists and gardeners alike to identify ornamentally planted trees, and to anticipate what any given specimen tree will look like if bought and planted. I hope that the range of popular garden species included in the book helps in this regard.

COLLINS
COMPLETE GUIDE TO
BRITISH TREES

Paul Sterry

WILLIAM
COLLINS

This book is dedicated to the memory of Bramley Frith.

HarperCollinsPublishers
1 London Bridge Street
London, SE1 9GF

www.harpercollins.co.uk

HarperCollinsPublishers
1st Floor, Watermarque Building, Ringsend Road
Dublin 4, Ireland

First published in 2007

ISBN: 978-0-00-723685-5

Colour reproduction by Nature Photographers Ltd.
Edited and designed by D & N Publishing, Hungerford, Berkshire
Printed and bound in Bosnia and Herzegovina by GPS Group

HOW TO USE THIS BOOK

The book has been designed so that the text and photographs for each species are on facing pages. A system of labelling clearly identifies each tree or shrub. The text complements the information conveyed by the photographs. By and large, the order in which species appear in the main section of the book roughly follows standard botanical classification. However, because parts of the field are in a state of flux, the order may differ slightly from that found in other guides, past, present or future.

SPECIES DESCRIPTIONS

At the start of each species description the most commonly used and current English name is given. My primary source of reference for this, and for scientific names, has been Clive Stace's *New Flora of the British Isles* (*see* p. 312). Essentially this covers native and widely naturalised British species. For non-native garden trees and shrubs I have followed the naming employed by Owen Johnson in Collins' *Tree Guide*.

The English name of each plant is followed by its scientific name, which comprises the genus name followed by the specific name. In a few instances, where this is pertinent, reference is made (either in the species heading or in the main body of the text) to a further subdivision: subspecies. Some cultivated trees and shrubs are known best, or sometimes exclusively, by their cultivar name, so where this helps with recognition I have included it; some hybrid trees and shrubs are now known *only* by their horticultural names.

The text has been written in as concise a manner as possible. Each description begins with a summary of the tree or shrub. To avoid potential ambiguities, the following subheadings break up the rest of each species description: BARK; BRANCHES (occasionally SHOOTS); LEAVES; REPRODUCTIVE PARTS; STATUS AND DISTRIBUTION; COMMENTS. Not all these headings are used for every species. Within the sections dealing with plant parts, colour, shape and size are described; these tend to be more constant, and hence are more useful for identification, than the overall size of the tree or shrub.

MAPS

The maps provide invaluable information about the distribution and occurrence of species in the region, but obviously they are only relevant in the case of native species and widely naturalised ones. Magenta has been used to represent the range of native species while cyan denotes the occurrence of naturalised alien trees or shrubs; the intensity of the colour gives an indication of a species' abundance in a given area. In compiling the maps, I have made reference to a number of sources, including *An Atlas of the Wild Flowers of Britain and Northern Europe*, various county floras, the *New Atlas of the British and Irish Flora* (*see* p. 312) and my own notes. The maps represent the current ranges of trees and shrubs in the region in general terms. Please bear in mind that, given the size of the maps, small and isolated populations will not necessarily be featured. Furthermore, the ranges of many species (particularly invasive introduced species and declining native trees and shrubs) are likely to change as the years go by.

PHOTOGRAPHS

Great care has gone into the selection of photographs for this book and in many cases the images have been taken specifically for this project. Preference was given to photographs that serve both to illustrate key identification features of a species and to emphasise its beauty. In many instances, smaller inset photographs illustrate features useful for identification that are not shown clearly by the main image.

Map coloration helps distinguish between the native and introduced distributions of species such as the Scots Pine.

GLOSSARY

Achene – one-seeded dry fruit that does not split.

Acute – sharply pointed.

Alien – introduced by man from another part of the world.

Alternate – not opposite.

Anther – pollen-bearing tip of the stamen.

Auricle – one of a pair of lobes at the base of a leaf.

Axil – angle between the upper surface of a leaf, or its stalk, and the stem on which it is carried.

Berry – fleshy, soft-coated fruit containing several seeds.

Bract – modified, often scale-like leaf found at the base of flower stalks in some species.

Calcareous – containing calcium, the source typically being chalk or limestone.

Calyx – outer part of a flower, comprising the sepals.

Capsule – dry fruit that splits to liberate its seeds.

Catkin – hanging spike of tiny flowers.

Clasping – descriptive of leaf bases that have backward-pointing lobes which wrap around the stem.

Compound – (of leaves) divided into a number of leaflets.

Cordate – heart-shaped at the base.

Corolla – the collective term for the petals.

Cultivar – plant variety created by cultivation.

Deciduous – plant whose leaves fall in autumn.

Dentate – toothed.

Dioecious – having male and female flowers on separate plants.

Drupe – succulent or spongy fruit, usually with a hard-coated single seed.

Entire – (of leaves) with an untoothed margin.

Fruits – the seeds of a plant and their associated structures.

Glabrous – lacking hairs.

Globose – spherical or globular.

Hybrid – plant derived from the cross-fertilisation of two different species.

Inflorescence – the flowering structure in its entirety, including bracts.

Introduced – not native to the region.

Involucre – ring of bracts surrounding a flower or flowers.

Lanceolate – narrow and lance-shaped.

Leaflet – leaf-like segment or lobe of a leaf.

Lenticel – breathing pore on a fruit, shoot or trunk.

Linear – slender and parallel-sided.

Lobe – a division of a leaf.

Midrib – the central vein of a leaf.

Native – occurring naturally in the region and not known to have been introduced.

Oblong – (of leaves) with sides at least partly parallel.

Obtuse – (of leaves) blunt-tipped.

Opposite – (usually of leaves) arising in opposite pairs on the stem.

Oval – leaf shape.

Ovary – structure containing the ovules, or immature seeds.

Ovoid – egg-shaped.

Palmate – (of leaves) with finger-like lobes arising from the same point.

Pedicel – stalk of an individual flower.

Perianth – collective name for a flower's petals and sepals.

Petals – inner segments of a flower, often colourful.

Petiole – leaf stalk.

Pinnate – (of leaves) with opposite pairs of leaflets and a terminal one.

Pod – elongated fruit, often almost cylindrical, seen in pea family members.

Pollen – tiny grains that contain male sex cells, produced by a flower's anthers.

Pubescent – with soft, downy hairs.

Rachis – main stalk of a compound leaf or stem of an inflorescence or array of fruits.

Reflexed – bent back at an angle of more than 90 degrees.

Sepal – outer, usually less colourful, segments of a flower.

Stamen – male part of the flower, comprising anther and filament.

Stigma – receptive surface of the female part of a flower, to which pollen adheres.

Style – an element of the female part of the flower, sitting on the ovary and supporting the stigma.

Tepal – perianth segment when petals and sepals are not identifiably separable.

Tomentose – covered in cottony hairs.

Whorl – several leaves or branches arising from the same point on a stem.

BASIC TREE BIOLOGY

In many ways, trees and shrubs are no different from other flowering plants – they just happen to be bigger. They all grow, produce leaves, flowers and fruits in order to reproduce, and compete with other forms of life in the struggle to survive. The following is a basic review of the biology of trees and shrubs, which will help any reader unfamiliar with the subject to gain a better understanding of these fascinating organisms.

WOODY TISSUE

The principal way in which a tree or shrub differs from other, herbaceous, members of its family is its ability to produce woody tissue; this serves to conduct materials around the plant, and leads to the production of permanent shoots. In the case of perennial herbaceous plants, the shoots die back at the end of each growing season, or in the case of annuals, the whole plant dies and a new generation arises from seeds formed by the previous generation.

Trees and shrubs have an important layer of cells enclosing shoots, buds and roots, called the *cambium* layer. This is an active layer that is constantly producing new cells on its inner and outer surfaces. Cells that grow on the inside of the cambium develop into woody tissue or *xylem*; this conducts water from the roots to the shoots, buds and leaves. Eventually it forms the bulk of the trunk and branches of the tree as a new layer is laid down each year. Cells that grow on the outside of the cambium form the conductive tissue, known as *phloem*, that carries sugars from the leaves down to the roots. This vital layer must not be damaged. If a complete ring of this tissue is cut away from the trunk of a tree the roots will be deprived of nourishment from the leaves and the tree will eventually die.

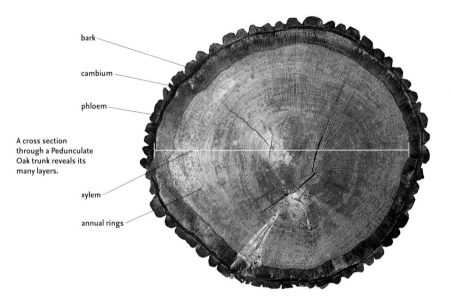

bark

cambium

phloem

A cross section through a Pedunculate Oak trunk reveals its many layers.

xylem

annual rings

ANNUAL RINGS

When the tree begins to grow vigorously in the spring it forms large conductive cells that allow sap to flow freely through the trunk. As the season advances, cells produced by the cambium layer become smaller, with thicker walls for support, so they give a more dense appearance. In winter, cell production slows down and then ceases for a while; come the spring, however, there is a sudden burst of cell production and large cells are produced once more. The new growth of large cells immediately next to the thinner layer of dense cells gives the appearance of a ring. By examining a cut stump it is possible to count the rings and therefore discover the age of the tree, and also to find out which were the best growing seasons (*see* pp. 24–5).

BARK

The bark is an important part of a tree, protecting the vital growing layers of cells below from varying environmental conditions. It is produced by a layer of cambium cells and grows to accommodate the increasing girth of the tree. It may be thin and papery, smooth and shiny, or thick and deeply furrowed. Each type of bark is matched to the tree's environment, so tree species that are subject to heat and strong sunlight in their native ranges have a thicker bark than those that come from cooler, humid climates (*see also* pp. 49–54).

Himalayan Birch bark (*left*) is relatively thin and peels readily, while Pyrenean Oak bark (*right*) is thick and corky.

ROOTS

The first part of a tree to emerge from a seed is a tiny root whose first function is to draw up water and dissolved minerals from the soil. In the case of most of our tree species, successful germination is dependent upon this first root making contact with a species-specific symbiotic fungal partner, a relationship that continues for the rest of the tree's life. This relationship is discussed in more detail on p. 34. From this simple start the tiny root will grow and divide, eventually forming a large network of powerful roots, side-branches and fine root hairs that spread out in all directions around the base of the trunk. The main roots will be woody and very strong, but their many branches terminate in fine root hairs that are only a few cells long; these have thin, permeable walls through which will pass all the water and minerals needed for the survival and growth of the tree. Although the sturdy roots strengthened with woody tissue help anchor the tree in the ground, it is the millions of fine root hairs that keep the tree alive by supplying it with water and nutrients. These fine root hairs are very short-lived, being constantly replaced as the main roots grow through the soil.

The root system of a large, mature tree does not penetrate far down into the soil. The most useful supply of dissolved nutrients for the tree lies in the shallow layer of topsoil and the adjacent sub-soil, so it is more beneficial if the roots spread outwards through this layer rather than penetrate to a great depth into a rather sterile and hostile layer. A 50m-tall tree will probably have a spread of smaller branching roots all around the bole, the extent approximately equal to the spread of the branches or, sometimes, to the height of the tree. The proximity of other trees, the nature of the soil, and the presence of obstacles like rocks or river banks will all influence the final extent of the root system, however. This knowledge of the spread of the roots is useful when planning where to plant large trees

Spreading Beech roots.

that may damage drains or the foundations of buildings when they reach maturity, and it should also be borne in mind when digging ditches or ponds near large trees.

In order to be able to function at all, roots require a supply of nutrients from the leaves, so within the root system there is a two-way traffic of water and minerals up from the soil to the leaves, and dissolved sugars and other nutrients down from the leaves.

The root hairs are living cells that require oxygen in order to be able to carry out respiration. They give off carbon dioxide as a waste gas, so they need access to air in the soil to allow these gases to circulate. Most trees, and most land plants, cannot grow in completely water-logged soils and those that do have special adaptations for survival.

A number of trees, especially members of the Fabaceae, such as the Honey Locust, have many rounded nodules on the roots that contain colonies of nitrogen-fixing bacteria. These are able to use gaseous nitrogen and turn it into compounds vital to the growth of the plants.

LEAVES

Leaves are among the most conspicuous and distinctive features of any tree. They grow in a huge variety of shapes, sizes, colours and combinations and are usually the best feature for identifying the tree because of their unique structure. Leaves may vary from one species to another but they all perform the same vital function as the principal producers of food for the tree.

The first pair of leaves to emerge from the seed are simple, and bear no resemblance to the true leaves of the tree; they are derived from the seed's food store. They are green, however, and supply the tiny seedling with its first food made from sunlight energy. Once the seedling has begun to produce leaves that are miniature versions of its true leaves, growth can begin very rapidly. Tiny seedlings are vulnerable to grazing, trampling, drought and competition, so very few survive.

Recently germinated oak seedling.

A plant's leaves are its powerhouse, trapping energy from sunlight and converting it into basic food.

Leaves are basically thin layers of living tissue with the ability to trap light energy and use this to convert the raw materials of water and carbon dioxide into a simple sugar. This reaction, known as *photosynthesis*, is arguably the most important chemical reaction in the world, for it is the basis of all other food production. Animals do not have the ability to convert these simple materials into food; they have to rely on plants to do it for them. The simple sugar produced in the leaves is glucose, and this can be formed into a variety of other important materials, particularly starch, which many plants store, or pack into their seeds. A vital by-product of this reaction is oxygen, the gas essential for the respiration of all members of the animal kingdom. This explains the vital role of trees in the ecosystem: they are major consumers of carbon dioxide, one of the so-called 'greenhouse gases'; and they are major producers of oxygen, the gas we need for our respiration. They are also major producers of food for much of the animal kingdom.

Contained within a leaf are numerous specialised cells. Some are concerned with the transport of materials in and out of the leaf, some are the vital energy-trapping cells that utilise sunlight, and others are concerned with the regulation of water movements. The cells that trap light-energy contain a light-absorbing pigment called *chlorophyll*, which gives leaves their green colour. Other pigments of different colours may be present in varying amounts, and it is this variety that gives leaves of different trees their own subtle shade of green. Without the green chlorophyll or other light-absorbing pigments, leaves would be unable to perform their important function, and also, if deprived of light, they would be unable to manufacture the tree's food.

Leaves arrange themselves in such a way to absorb the maximum amount of sunlight, so spreading canopies or trees growing taller than their neighbours, are both ways in which trees maximise the light-gathering power of their leaves. Some leaves have paler patches that lack green chlorophyll; these are known as variegated leaves and certain trees, such as some cultivars of the Highclere hollies, regularly produce green-and-yellow leaves. If the leaves were completely lacking in chlorophyll they would be unable to manufacture food for the tree; the small areas of green tissue in the leaf produce all the food needed by the whole leaf.

All leaves have tiny pores in their surface (normally just the lower surface) called *stomata*. These allow water to evaporate into the atmosphere. To some extent the tree can regulate the opening and closing of these stomata, but during daylight hours, when the tree is trapping sunlight, they will be open, allowing water out and also allowing the circulation of the gases involved in photosynthesis. This can lead to problems for trees growing in hot, dry areas, or in well-drained soils where little ground water is available. In order to allow the essential gases to circulate, and at the same time minimising water loss, many leaves have become reduced in size, such as the needles of firs and pines, or have thick waxy upper surfaces such as the glossy green leaves of hollies and magnolias. This reduces water loss to a minimum without impeding photosynthesis.

The great variety of leaf shapes and sizes is an indication of the variety of ways in which trees can cope with environmental conditions. Some trees grow in areas where water is at

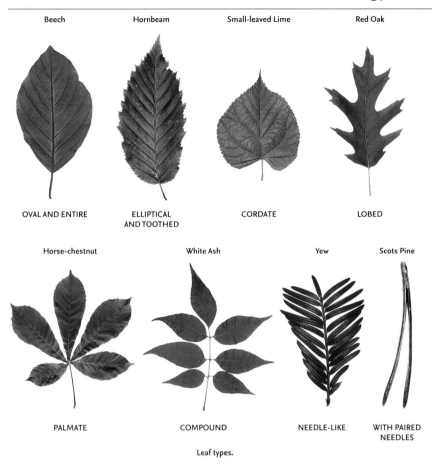

Beech	Hornbeam	Small-leaved Lime	Red Oak
OVAL AND ENTIRE	ELLIPTICAL AND TOOTHED	CORDATE	LOBED

Horse-chestnut	White Ash	Yew	Scots Pine
PALMATE	COMPOUND	NEEDLE-LIKE	WITH PAIRED NEEDLES

Leaf types.

a premium, so they have small leaves, to cut down on water loss through their thin skins. Some grow in shady conditions, so they may have larger leaves that can trap the maximum amount of light energy. Some trees are subject to grazing by animals, so their leaves are spiny or prickly, or protected on tough, thorny stems.

Deciduous trees, such as Horse-chestnut, produce fresh leaves each spring, which burst forth from buds.

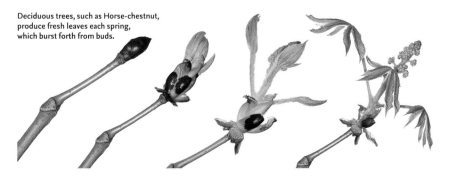

Autumn leaf colour is spectacular in many maple species: as chlorophyll
and other pigments are withdrawn, remaining red pigments prevail.

Evergreen trees do not lose all their leaves at the end of every growing season; most leaves remain on the tree through the winter, although there is always some loss and some replacement. In many of the pines, for example, the needles will remain on the tree for about 3 years. As the shoot grows longer each year, a new set of needles grows on the tip of the lengthening shoot. The older needles, finding themselves further and further away from the tip, gradually fall off. Small leaf scars remain, and these are quite distinctive in some species and may be a useful aid to identification. Broadleaved trees such as Holly also replace their leaves gradually so there is always some leaf-fall, but plenty of green foliage remains on the tree.

Deciduous trees generally shed all their leaves at the end of the growing season, before the onset of winter. Many of them produce spectacular displays of colour before the leaves finally fall. These colour changes are the result of the gradual withdrawal back into the tree of all the useful materials in the leaf; as the various pigments are removed the leaf itself changes colour until finally a corky layer, called the abscission layer, grows at the base of the petiole or leaf stalk. This seals off the shoot and when the leaf finally falls, a scar is left through which mould spores and other harmful materials are unable to pass. The twigs of Horse-chestnut have very distinctive leaf scars that look like tiny horseshoes. If these are examined carefully through a hand-lens, the sealed-off ends of the vessels that conducted materials in and out of the leaves can clearly be seen.

There may be as many as 250,000 leaves on a mature oak tree, whilst a large spruce probably has 10 times as many, in the form of needles. The oak's leaves will be shed at the end of the growing season, adding to the rich accumulation on the ground beneath it, whilst the spruce's needles will be shed and replaced gradually, each individual needle remaining on the tree for about 4 years.

REPRODUCTION

Trees normally produce flowers when they are several years old. Beech, for example, produces its first flowers at around 30 years old, repeating the process each spring for the next 200 years if it remains healthy. Some trees, such as apples or oaks, have years in which they produce a large crop of fruits or seeds, followed by other years with hardly any, whilst other species, such as some maples, produce a good seed crop year after year.

Even a slight breeze will liberate pollen from the male catkins of Hazel (*right*) and carry it to female flowers (*far right*).

Some trees and shrubs produce conspicuous flowers to attract pollinating insects, something that, in ornamental trees, we also find attractive. Honey Bees are particularly important pollinators, but numerous other insects visit the flowers for nectar and pollen. Many flowering trees have also long been prized by gardeners for their scent.

Many trees are pollinated by the wind. Their flowers are less conspicuous, often taking the form of catkins, which are pendulous and usually open before the leaves so that nothing impedes the free movement of the pollen grains. Wind-pollinated flowers normally have flowers of separate sexes, the males usually being larger and more abundant. Many wind-pollinated trees are such prolific producers of pollen that on warm breezy days in spring clouds of pollen can sometimes be seen blowing from the trees.

Conifer flowers are either male or female, and borne on the same or different trees. There are no petals, but some of the flowers are still quite colourful and decorative. Male flowers are short-lived, falling off after they have released clouds of pollen, but the female flowers, often covered with brightly coloured scales, remain on the tree after pollination and develop into cones containing the seeds. They rely on the wind for pollination and also for seed dispersal. A few close relatives of the conifers, such as the yews, produce fleshy fruits instead of cones.

Mature cones open and close in response to temperature and humidity, releasing seeds in hot, dry conditions.

The flowers of broadleaved trees and shrubs are usually hermaphrodite, containing both male and female parts, but there are a number of exceptions. Both sexes usually have petals in some form or other and they may also be scented. Small flowers are often grouped together in larger clusters to help attract pollinating insects. Some are wind-pollinated and open early in the year before the leaves, but insect-pollinated flowers usually open in spring and summer when more insects are active.

The fruits of trees and shrubs are much more varied than the cones of the conifers. They range from tiny papery seeds with wings, through nuts and berries, to large succulent fruits in a variety of shapes and colours. Edible fruits are designed to assist dispersal of the seeds by animals and many are delicious to the human palate.

Hawthorn flowers are an attractive sight in spring and are irresistible to pollinating insects.

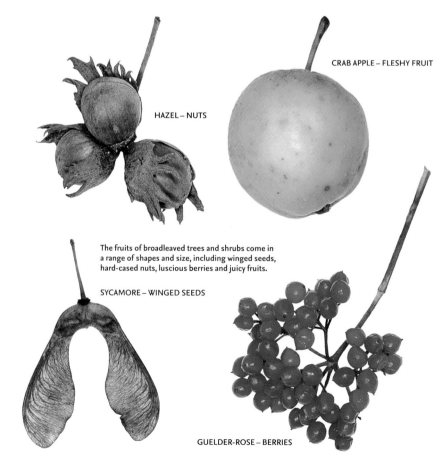

CRAB APPLE – FLESHY FRUIT

HAZEL – NUTS

The fruits of broadleaved trees and shrubs come in a range of shapes and size, including winged seeds, hard-cased nuts, luscious berries and juicy fruits.

SYCAMORE – WINGED SEEDS

GUELDER-ROSE – BERRIES

WHAT IS A TREE?

Nobody could have any doubt that a mature Wellingtonia or an ancient, spreading Pedunculate Oak is a tree, but would a prostrate Juniper, or a Creeping Willow, also qualify, or are they merely shrubs? One feature common to both trees and shrubs is that their stems increase in thickness each year by the laying down of internal layers of woody tissue in the form of concentric rings. This secondary thickening builds up year by year to increase the diameter of the stem and gives a permanent record of the age of the tree or shrub.

Trees are generally considered to have a single main stem of 5m or more in height with a branching crown above this, whereas shrubs may have numerous stems arising at ground level and may not normally reach the height of a tree. Both trees and sizeable shrubs are covered in this book but the distinction is not always clear. Individuals of the same species may become trees or form shrubs, depending on the circumstances in which they are growing, or their management. Hazel, for example, forms a multi-stemmed shrub in response to coppicing, or cutting back to the rootstock; each time this is done, new shoots arise from the rootstock and the Hazel regenerates. If this cutting back does not take place, it can grow as a medium-sized tree on a single stem.

Trees do not belong to a single family of plants; many plant families are represented, and some, like the Fabaceae and Rosaceae, also include many herbaceous plants and shrubs as well as large trees.

The plant kingdom is divided into two main classes, the Gymnosperms and the Angiosperms. The most primitive of the two classes is the Gymnosperms, the name meaning 'naked seeds'; the ovule is borne on a bract and not enclosed in a seedpod or case. This class includes the Maidenhair Tree, a very primitive tree, and all the conifers, or cone-bearing trees.

The name Angiosperm means 'hidden seeds' and refers to the way the seeds are contained inside an ovary, a structure that may later develop into a seedpod or fruit. This large class includes many well-known plant families, some of which are mostly made up of herbaceous plants, and some of which are mostly composed of trees. The Limes (Tiliaceae) and Elms (Ulmaceae), for example, are mostly trees, whilst trees feature in only one of many genera in the Foxglove family (Scrophulariaceae). All of our garden and wild flowers, bulbs, palms and the grasses and cereals are Angiosperms.

With their massive trunks, there is no mistaking that these Common Beeches are indeed trees.

TREE AND SHRUB FAMILIES IN BRITAIN AND IRELAND

GINKGOACEAE (GINKGO FAMILY) – *see* p. 68
An ancient family, representing the precursors of our modern conifers and broadleaves, that thrived before present tree families had evolved. Only one species has survived; others are known only from fossils dating from at least 200 million years ago.

ARAUCARIACEAE (MONKEY-PUZZLE FAMILY) – *see* p. 68
A family of large evergreen trees, some important timber-producers, found mainly in South America and Australasia. Sexes are separate and trees seen in Britain have been raised from seed.

TAXACEAE (YEW FAMILY) – *see* p. 70–2
A small family of primitive conifers, some being little more than shrubs, restricted to the northern hemisphere. They have poisonous seeds and foliage. Male and female flowers are produced on separate trees and the seeds are surrounded by a fleshy cup called an aril. They can be propagated by seeds and cuttings.

CEPHALOTAXACEAE (PLUM YEW or COW-TAIL PINE FAMILY) – *see* p. 72
Once a widely distributed family, according to fossil remains, but now restricted to the Far East. The leaves are large, flattened needles, the male and female flowers are borne on separate plants and the fruits are plum-like.

PODOCARPACEAE (PODOCARP or YELLOW-WOOD FAMILY) – *see* p. 74
A family of yew-like trees with fruits that are borne on fleshy stalks that are edible. Mainly confined to the tropics and the southern hemisphere, but some occur in Japan and India.

CUPRESSACEAE (CYPRESS FAMILY) – *see* p. 76–90
A large group of coniferous trees widely spread around the world. Most have very small, scale-like leaves and tiny buds. The cones are small and tough, often rounded and woody, or fleshy in the case of junipers. Most are slow-growing and long-lived, giving strong, scented timber, and were mistakenly called cedars by early explorers.

TAXODIACEAE (REDWOOD FAMILY) – *see* p. 92–6
An ancient family, once with many more representatives than the 15 species that exist today. Four species are deciduous, the others all evergreen and mostly with hard, spine-tipped leaves. The globular cones are relatively small. The bark in all species is fibrous and a rich red-brown. Some redwoods are the largest living organisms in the world.

PINACEAE (PINE FAMILY) – *see* p. 98–128
A large family of 200 species, all originating in the northern hemisphere. Their cones are woody and composed of a spiral arrangement of scales, each with two seeds. The leaves are needle-like. The arrangement of the needles, such as being grouped in pairs or threes, or growing on short pegs, is a great help in the identification of these trees. The family includes firs (genus *Abies*), cedars (genus *Cedrus*), larches (genus *Larix*), spruces (genus *Picea*), hemlock-spruces (genus *Tsuga*), Douglas firs (genus *Pseudotsuga*) and pines (genus *Pinus*).

SALICACEAE (WILLOW AND POPLAR FAMILY) – *see* p. 130–46
A very widespread group of trees and shrubs, numbering well over 300 species, with many more hybrids that are often difficult to place. Male and female flowers are found on separate trees and usually take the form of catkins. Leaves are alternate, and long and pointed in the case of willows. Most species, apart from Goat Willow and Grey Willow, propagate easily from cuttings. Many grow in wet habitats such as stream sides, and small shrubby species are often found in upland regions. Wind-dispersed seeds make them rapid colonisers of new habitats. Most members of this family are vigorous, fast-growing trees and tolerant of much bad treatment from both man and natural disasters like storms. Many are of great importance for wildlife, supporting large numbers of insect larvae. A number are grown for ornament, having a weeping habit, bright foliage or colourful winter twigs.

JUGLANDACEAE (WALNUT FAMILY) – *see* p. 148–50
A family of 7 genera and about 60 species spread across the Americas, SE Europe, SE Asia and Japan. Leaves are usually alternate and pinnate, flowers are without petals, small and grouped in catkins, with males and females on the same plant. The fruit is usually a nut, sometimes large and edible, or sometimes small and winged. The family includes hickories (genus *Carya*), wingnuts (genus *Pterocarya*) and walnuts (genus *Juglans*).

MYRICACEAE (BOG MYRTLE FAMILY) – *see* p. 150
A family of 2 genera and 35 species of trees and shrubs; only one in our region. The simple leaves have resinous glands and the flowers are borne in spike-like catkins.

BETULACEAE (BIRCH FAMILY) – *see* p. 152–64
A large family of 6 genera and about 150 species of medium-sized trees and shrubs. Flowers are in the form of catkins, with the separate sexes growing on the same tree; the male catkins are the more conspicuous. Seeds are borne in smaller cone-like catkins, or in the form of nuts with hard shells or sometimes wings. In the British Isles, the family is represented by birches (genus *Betula*), alders (genus *Alnus*), hornbeams (genus *Carpinus*), hop-hornbeams (genus *Ostrya*) and hazels (genus *Corylus*). Some authorities place the genera *Carpinus* and *Ostrya* in a separate family, Carpinaceae, while the genus *Corylus* is sometimes placed in the family Corylaceae.

FAGACEAE (BEECH FAMILY) – *see* p. 166–80
A large family containing many well-known trees. More than 1,000 species, in 8 genera, occur, mostly in the northern hemisphere, but many far to the south. The flowers are small, sexes are usually separate and on the same tree. Fruits are in the form of nuts, protected by a cupule. The family is represented in Britain and Ireland by beeches (genus *Fagus*), southern beeches (genus *Nothofagus*), Sweet Chestnut (genus *Castanea*) and oaks (genus *Quercus*).

ULMACEAE (ELM FAMILY) – *see* p. 182–8
Includes about 150 species of both deciduous and evergreen trees and shrubs occurring in tropical and northern areas. The leaves are normally alternate, the small flowers lack petals and the fruits may be winged, in the form of a nut, or fleshy with a single stone.

MORACEAE (MULBERRY FAMILY) – *see* p. 190
A large family from the tropics with around 1,000 species, 12 of which are known as mulberries. Two of these are hardy in Britain and Ireland; their male and female flowers are in the form of separate catkins, but growing on the same tree, and the fruits are edible berries.

BERBERIDACEAE (BARBERRY FAMILY) – *see* p. 190
A family of shrubs with alternate leaves and flowers with 6–9 perianth segments in whorls. Fruits are berries or capsules. Only one species is native to the British Isles; it seldom achieves great stature.

PROTEACEAE (PROTEA FAMILY) – *see* p. 192
A large family of over 1,000 evergreen trees and shrubs, mostly native to the southern hemisphere, but introduced widely around the world. Leaves are alternate and sometimes pinnate. The flowers can be very showy, although petals are very small, the main display being provided by a large divided calyx.

CERCIDIPHYLLACEAE (KATSURA FAMILY) – *see* p. 192
A very small family with probably only a single species, once thought to be closer to the magnolias, but now considered to be more primitive and perhaps nearer to the planes.

MAGNOLIACEAE (MAGNOLIA FAMILY) – *see* p. 192–4
A family of 12 genera and up to 200 species, most occurring in Asia, particularly the Himalayas, China and Japan; a few also occur in the south of the USA and further south into South America. They can be either trees or shrubs, deciduous or evergreen, with alternate, untoothed and occasionally lobed leaves. The flowers are often showy and sometimes scented.

LAURACEAE (LAUREL FAMILY) – *see* p. 196
Mostly evergreen trees and shrubs. The family numbers about 1,000 species, mainly found in the tropics, but with a few hardy species occurring in more northern areas. Many are aromatic.

HAMAMELIDACEAE (WITCH HAZEL FAMILY) – *see* p. 198
Contains about 25 genera and 100 species that occur in temperate and subtropical regions. They range from trees to shrubs and may be evergreen or deciduous. Many are very popular ornamental garden plants.

PLATANACEAE (PLANE FAMILY) – *see* p. 200
A family of 8 species of large deciduous trees, mostly native to the USA and Mexico, apart from one that occurs in the Balkans and one in SE Asia. Leaves are large and normally palmate. Male and female flowers are in separate pendulous clusters on the same tree.

ROSACEAE (ROSE FAMILY) – *see* p. 202–56
A very large and important family of over 100 genera and about 3,000 species of trees, shrubs and herbaceous plants. The trees can be deciduous or evergreen, and have alternate, simple leaves or a range of leaf types including complex pinnate leaves. Flowers are usually 5-petalled, with the ovary beneath the petals, but the fruits are very varied and the family is divided mainly on the basis of the types of fruits produced. Tree and shrub representatives in Britain and Ireland are brideworts (*Spiraea*), Quince (*Cydonia*), Medlar (*Mespilus*), pears (*Pyrus*), apples (*Malus*), whitebeams and allies (*Sorbus*), Loquat (*Eriobotrya*), mespils (*Amelanchier*), cotoneasters (*Cotoneaster*), hawthorns (*Crataegus*) and cherries and their allies (*Prunus*).

FABACEAE (PEA FAMILY) – *see* p. 258–60
A very large family of trees, shrubs and herbaceous plants that bear their seeds in pods. Their roots have colonies of nitrogen-fixing bacteria living on them in tiny nodules. All of the tree species have tough, durable wood and many of them are thorny. Most have compound leaves and very attractive flowers.

SIMAROUBACEAE (QUASSIA FAMILY) – *see* p. 260
A mainly tropical and subtropical family of about 20 genera and 150 species of trees and shrubs. Leaves are alternate and usually pinnate, and the flowers are small and 5-petalled. The fruit is either winged or a capsule.

ANACARDIACEAE (CASHEW FAMILY) – *see* p. 260
A large family of more than 800 species of trees, shrubs and climbers found mainly in warm climates. They may be deciduous or evergreen, but most have alternate leaves that can be simple or pinnate, and many, such as Poison Ivy, have an irritant resin in the leaves that can damage human skin. The flowers are small and the sexes are often on different plants.

ACERACEAE (MAPLE FAMILY) – *see* p. 262–72
A family of about 100 species of trees and shrubs, some evergreen, some deciduous, mostly occurring in northern temperate regions. Leaves are opposite and nearly always lobed, and sometimes divided into leaflets. Flowers are small, and the seeds are winged, in 2 halves. Many have beautiful autumn colours and are popular garden trees, and some are important timber-producing trees.

HIPPOCASTANACEAE (HORSE-CHESTNUT FAMILY) – *see* p. 274–6
A family of 2 genera and 15 species of deciduous trees and shrubs occurring in North America, SE Europe and E Asia. The compound leaves are strongly palmate, and the showy flowers are 4- or 5-petalled, growing in large upright clusters at the ends of the shoots, usually in summer. The fruits are large shiny nuts in a variably prickly husk. The timber is not particularly strong for such a large tree, and is best used for carving and turnery.

AQUIFOLIACEAE (HOLLY FAMILY) – *see* p. 278
A large family of evergreen and deciduous trees and shrubs from temperate and tropical regions, most of which are hollies. Leaves are usually alternate. The male and female flowers are normally small, white or tinged pink, and on separate plants. The fruit is usually a colourful berry.

Celastraceae (Spindle family) – *see* p. 280
A family of almost 100 genera and more than 1,000 species of evergreen and deciduous trees, shrubs and climbers, found in many parts of the world and in many climatic types. The leaves may be opposite or alternate and the greenish flowers are usually small and insignificant.

Buxaceae (Box family) – *see* p. 280
A family of about 60 species of evergreen trees and shrubs, with a few herbaceous plants. The leaves are normally opposite and the flowers are tiny, usually growing in clusters.

Rhamnaceae (Buckthorn family) – *see* p. 282
A family of about 60 genera and 900 species of trees, shrubs and climbers found in most regions of the world. They may be deciduous or evergreen, bear spines on the shoots and branches and have alternate or opposite leaves. The flowers are small and separate sexes may occur on different plants. A number of species yield useful dyes, and many are poisonous.

Tiliaceae (Lime family) – *see* p. 284–6
A family of more than 700 species of trees, shrubs and herbaceous plants. The majority are found in the tropics but the 30 true limes (*Tilia*), which originated in the cooler regions of the northern hemisphere, are the only trees. The leaves are alternate and may be lobed, and they often have star-like hairs. The flowers are small, frequently fragrant, with 5 petals and sepals and many stamens. The fruit is usually a dry capsule, but it may be hard and woody. The timber is pale and soft and can be used for wood-carving.

Pittosporaceae (Pittosporum family) – *see* p. 288
A large family of 9 genera and over 200 species, mostly originating in Australasia. The leaves are alternate and usually untoothed and the 5-lobed flowers develop into either dry or succulent fruits. Some are cultivated and numerous attractive varieties occur in gardens.

Tamaricaceae (Tamarisk family) – *see* p. 288
A family of small trees and shrubs with tiny, scale-like, clasping alternate leaves and glands that excrete salt. The flowers are small but borne in dense heads. The seeds are wind-dispersed and are good colonisers of disturbed ground. Many grow near the sea.

Elaeagnaceae (Oleaster family) – *see* p. 288
A family of about 50 species of evergreen or deciduous shrubs and small trees, found in temperate and cooler regions in the northern hemisphere. The twigs are frequently armed with spines. The leaves have entire margins, may be scaly on the underside and are either opposite or alternate. The flowers are small and lack petals, and the sexes may be separated on different plants. Some species produce edible fruits.

Myrtaceae (Myrtle family) – *see* p. 290–2
A large family mainly occurring in the southern hemisphere. Only a single representative occurs naturally in Europe (the evergreen shrub Myrtle *Myrtus communis*), and none is known in North America. The family does, however, include the genus *Eucalyptus*, trees from Australasia. Overall, the family includes about 4,000 species of mostly evergreen aromatic trees and shrubs. The leaves are generally opposite and the flowers have 4 or 5 petals but many stamens.

Eucryphiaceae (Eucryphia family) – *see* p. 292
A small family of 5 species in a single genus, but widespread, occurring in Chile and Australasia. The leaves are opposite and may be simple or pinnate, with entire or toothed margins. One is deciduous but the others are evergreen. The flowers, borne in summer, are showy and make these trees popular subjects for large gardens.

Nyssaceae (Black Gum family) – *see* p. 292
A small family of 7 trees native to E Asia and North America. The male and female flowers are separate but borne on the same tree. The flowers are small and lack petals, but may be conspicuous because of large showy bracts below them. The leaves are alternate.

CORNACEAE (DOGWOOD FAMILY) – *see* p. 292–4
A family of about 100 species of evergreen or deciduous shrubs that grow in temperate regions of the northern hemisphere. Leaves are usually opposite and flowers are small but often surrounded by conspicuous colourful bracts.

ERICACEAE (HEATHER FAMILY) – *see* p. 296
A large family of 100 genera and about 3,000 species found all around the world. Most are trees or shrubs and may be evergreen or deciduous with alternate leaves. The flowers are variable, but most have 5 petals joined at the base.

EBENACEAE (EBONY FAMILY) – *see* p. 296
A large family of trees found mainly in the tropics and including the African tree that yields the black wood known as ebony. The date plums, *Diospyros* species, producing edible fruits, are mostly hardy in the N European climate.

STYRACACEAE (STORAX FAMILY) – *see* p. 296
A family of about 150 species of small trees and shrubs found in E Asia, the S USA, Central and South America, with an isolated species in the Mediterranean. The leaves are simple and alternate, and the flower has a tubular corolla dividing into 5–7 lobes. The fruit is a dryish capsule.

OLEACEAE (OLIVE FAMILY) – *see* p. 298–302
A large family of nearly 1,000 species, including many sweet-scented flowering shrubs such as lilacs (*Syringa*), and numerous large trees like ash (*Fraxinus*). All trees and shrubs have opposite leaves and the flowers have either 4 petals or none at all.

BIGNONIACEAE (BIGNONIA FAMILY) – *see* p. 304
A family of many trailing and climbing plants, but the *Catalpa* genus from North America, the West Indies and China contains 11 tree species.

SCROPHULARIACEAE (FIGWORT FAMILY) – *see* p. 304
A large family of mainly herbaceous plants such as the familiar foxgloves and speedwells. However, there are about 10 tree species, mainly from China, in the genus *Paulownia*.

CAPRIFOLIACEAE (HONEYSUCKLE FAMILY) – *see* p. 306–8
A family of shrubs and small trees, many with attractive flowers and colourful berries. Found across a range of habitats. A number of species have been taken into cultivation for food or as ornamental plants.

ASTERACEAE (DAISY FAMILY) – *see* p. 310
A large family, most members of which growing in the British Isles are annuals or perennial herbaceous plants. However, a few species do become woody shrubs. Flowers are typically numerous and borne in dense, terminal heads.

AGAVACEAE (CENTURYPLANT FAMILY) – *see* p. 310
A family of perennials with either succulent leaves in huge rosettes, or branched and woody stems that terminate in rosette-like tufts of leaves. Flowers are borne in large spikes. Mainly from warm climates or the southern hemisphere. Grown in the British Isles, usually in mild regions, and only one species really qualifies for tree-like status.

ARECACEAE (PALM FAMILY) – *see* p. 310
A large family from the tropics, adapted to survive in dry weather and often very windy conditions. The unusual form of growth does not enable them to branch, and the trunks do not become thicker as the trees age. Palms have a single terminal growing point from which new leaves and flowers emerge. Some are hardy enough to survive the European climate, growing in southern areas of Britain. Palms are raised from seed.

IDENTIFYING TREES

Very little equipment is required to identify trees, but a few simple items are helpful. The first essential is a notebook in which to record observations in the field. The traditional botanists' rule 'Take the book to the plant and not the plant to the book' applies to trees in that snapping off twigs, fruits or flowers should be avoided where possible; this may be prohibited, or may simply be bad manners if the tree in question is growing in a garden or park, for example. Some arboreta and botanical gardens are very firm in prohibiting the removal of living materials and may take action against a person who damages a tree. It is best if detailed notes can be made in the field, and if there are some fallen leaves or fruits beneath the tree, and it is certain that they are from the tree in question and not any other nearby trees, then these can be taken for help in identification.

Photographs are a very useful way of keeping records and will be of great help in identification. A record of the same tree in different seasons of the year would be helpful to show the changes in foliage colour and the appearance of the flowers.

A hand-lens, capable of magnifying up to × 10, is very helpful. Many features of trees, even very large specimens, are easier to see in close-up. Coniferous trees often have lines of tiny pores (stomata) on their needles, and the number and arrangement of these can be crucial to correct identification. The degree of hairiness of the leaves and petioles is often very important; some species have hairs in the form of tiny stars, while others may have fine hairs of a particular colour, both features that can only be reliably seen in close-up. The arrangement of the parts of the flowers often helps in tracing a tree to a particular family and these parts are usually quite small.

A small ruler with a scale marked in millimetres is also very useful as the measurements of needles, leaves, petioles and floral parts can be very important, although it should be stressed that these are often variable. A longer tape-measure is also helpful for measuring the circumference of trees; this information can be used to estimate the age of most trees (*see* p. 24).

ESTIMATING THE HEIGHT OF A TREE

Although the height of a tree is not a definitive guide to identification – trees grow after all – it often helps to know this measurement when deciding on a species' identity. There are a number of simple methods that can be used with reasonable accuracy but the following is the most straightforward.

1 Hold a straight stick vertically in front of you, at arm's length, and line it up with the tree. Move backwards or forwards until the stick appears to be the same height as the tree and then rotate the stick through 90 degrees until it is horizontal. **2** Keeping one end of the stick in line with the base of the trunk, ask a friend to stand in line with the other end of the stick, along the trajectory if the tree fell over and lay at right angles to your orientation. Measure the distance between your friend and the base of the tree trunk and this will tell you the height of the tree.

Measuring tree height.

THE LIFE CYCLE OF A TREE

Like all living things, trees have a finite life. Some live just a few decades while others may survive for hundreds of years; a few venerable specimens can claim to have seen out several millennia. However, whatever the life expectancy of a tree, it will pass through a series of reasonably well-defined stages, in much the same way as do all other flowering plants.

A tree's life begins when a seed germinates. The first pair of leaves to emerge is simple and they are quite unlike the true leaves of the tree or shrub; they are derived from the cotyledons that contain the seed's food store. These first leaves contain chlorophyll and can photosynthesise, supplying the tiny seedling with its first food made from sunlight energy. Tiny seedlings are vulnerable to environmental pressures – grazing, drought and trampling for example – and few make it to the next stage in a tree's life.

Many acorns are collected and buried by Jays. Those that escape being dug up will germinate the following spring.

Recently germinated seedling.

Vigorous sapling.

Soon the seedling will start to produce true leaves and once these appear then rapid growth begins and continues for many years. In the case of a Pedunculate Oak, for example, all being well it could put on up to half a metre of growth each year, both in terms of height and, more significantly, in terms of crown size. Of course, the amount of growth is influenced by environmental factors, including soil type, grazing and insect damage, the proximity of neighbouring trees and prevailing climatic conditions in any given year.

Many large native tree species will not produce flowers and fruits in any quantity, or at all, until they are at least 30 years old. After that, they continue to grow for several decades more, but a typical Pedunculate Oak is likely to reach pretty much its maximum size at around 100 years old. All being well, it will continue to produce acorns for at least the next 200 years, with the crown filling out and the trunk expanding – unless, of course, it is struck by environmental catastrophe, disease or the actions of man.

Mature oak.

Fallen giant.

During their heyday years, native deciduous tree species follow, and indeed mirror, the four seasons experienced by Britain and Ireland. They are leafless, and seemingly lifeless, during the winter months, but the sap begins flowing in spring and soon fresh leaves appear, followed by flowers. The leaves mature, and fruits form, in summer, the foliage losing some of the fresh green colour it had in spring. With the approach of autumn, nutrients are withdrawn from the leaves, resulting in colour change, and eventually in late autumn they fall, carpeting the ground below.

The four seasons.

In the natural course of events, all trees eventually come to the end of their lives, although in the case of the Pedunculate Oak and several other native species, this can be after several hundred years of existence. However, sooner or later, often as the result of disease and damage, the tree begins to die back. The appearance of fungal fruiting bodies on the trunk is often a sign that all is not well and that the process of decline and decay has begun. However, from an ecological perspective, there is something life-affirming about this process. In the absence of man's intervention, nothing goes to waste and all the nutrients from the venerable old tree enrich the soil, thanks to insect attack and fungal decay, thus creating ideal conditions for the next generation of trees. It should be borne in mind that the life of certain tree species can be prolonged by the actions of man. Notable coppiced Ash stools, for example, date back 800 years or more, much longer than an Ash tree would be likely to survive in its natural state.

GROWTH RINGS AND AGEING A TREE IN THE FIELD

For each year of its life a tree adds an internal ring of woody tissue to its trunk (*see* p. 8) and this leaves a permanent record of its age. Obviously, this record can be examined only if the tree is felled, or in some cases, if a fine core can be extracted from the bole, revealing a radial section of woody tissue. In some tree species growth rings can be difficult to discern and truly accurate ageing is best left to an expert dendrochronologist. However, ageing the cut stump of an oak, for example, is certainly not beyond the scope of the amateur naturalist, and a reasonable degree of accuracy can always be achieved.

Studying growth rings also provides a fascinating insight into the world of trees. In poor seasons fewer cells will be produced, so the final ring will not be as thick as one for a good growing season. This is a surprisingly accurate test of environmental conditions, as years in which there was known to be a summer drought, for example, can be shown to have narrower rings than warm and wet summers.

Examining cut stumps after forestry operations, measuring the circumference and counting the annual growth rings will give an idea of the age of trees of the same species of similar circumference. Other clues may help you to arrive at a reasonably accurate estimate of the tree's age. Many conifers, like the larches, for example, produce a new set of shoots around the crown each year while, lower down, the oldest branches die off and fall away, leaving stumps around the bole. Each year's growth produces a new ring of shoots around the main bole with a gap between them and the preceding year's shoots. If the number of rings of old branch stumps low down is added to the number of rings of growing branches higher up, the sum will give a rough guide to the age of the tree.

The circumference of the bole does give a good guide to the tree's age; however, for every species there is a point at which it reaches its optimum size. After that, it increases very slowly, if at all, gradually declining into old age. During the period of the tree's maturity, however, it is fairly safe to assume that each 2.5cm of circumference represents one year of the tree's life. This applies to normal healthy trees growing in places where they are free to attain their natural spread and height, unimpeded by other, neighbouring trees. It applies to most species of broadleaves and conifers growing in the open. Trees growing in woodland, with many other trees close by, will grow at only about half this rate, so before ageing a tree in this way, look at its surroundings, and at the crown and spread of the tree. Does it have a good-sized crown, or does it appear to be restricted in any way by other trees? If it appears to be growing normally with a good, healthy crown, then the 2.5cm-per-year rule can be applied.

Some trees are very slow-growing, so this rule cannot be applied to them. Yew, for example, grows at a rather pedestrian rate but is long-lived; in its old age it is so slow-growing that it hardly appears to increase in circumference at all. When using the circumference method of ageing a tree it is important to look for a vigorous and healthy tree in natural surroundings.

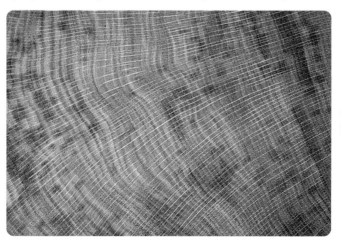

Close-up of annual growth rings of a Pedunculate Oak.

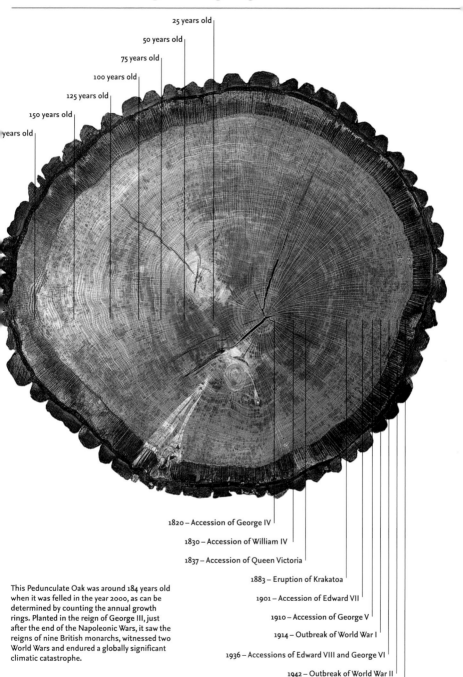

25 years old

50 years old

75 years old

100 years old

125 years old

150 years old

years old

1820 – Accession of George IV

1830 – Accession of William IV

1837 – Accession of Queen Victoria

1883 – Eruption of Krakatoa

1901 – Accession of Edward VII

1910 – Accession of George V

1914 – Outbreak of World War I

1936 – Accessions of Edward VIII and George VI

1942 – Outbreak of World War II

1952 – Accession of Queen Elizabeth II

This Pedunculate Oak was around 184 years old when it was felled in the year 2000, as can be determined by counting the annual growth rings. Planted in the reign of George III, just after the end of the Napoleonic Wars, it saw the reigns of nine British monarchs, witnessed two World Wars and endured a globally significant climatic catastrophe.

THE ECOLOGY OF TREES AND WOODLAND

Almost all the oak trees in Bramley Frith, Hampshire, are just a few hundred years old; much older ones were blown down in the Great Storm of 1703. Nevertheless, the woodland has had continuous tree cover since at least the Domesday Book.

Although there are many tracts of forest in the British Isles that harbour native species and ancient trees, it is important to realise that almost no woodland in Britain and Ireland can be described as truly virgin and untouched. For millennia man has interfered with the forested landscape, cutting down trees for fuel and building materials, and in order to create agricultural land. Some types of woodland are more modified than others and a few exist entirely as a result of human actions. However, when it comes to woodland comprising native species, the history of use does not necessarily detract from its ecological importance and value to wildlife; on the contrary, it often enhances it. Of greater significance than the age of the trees in a given area is a continuity of woodland cover. Many woodlands have had continuous tree cover since before man first settled the land and the diversity of associated wildlife reflects this venerable ancestry. The fact that ecologists refer to most British and Irish woodland as *semi-natural* is in no way derogatory.

Even as solitary individuals, native tree species will harbour a good array of wildlife. However, it is when they grow alongside and among other trees, and form woodland, that their true ecological potential becomes apparent. This can be thought of as synergy, if you like: from a wildlife perspective the whole (the woodland) is greater than the sum of the parts (the trees).

Walk through an area of recently planted woodland and you will usually discover a mixture of tree species, comprising individuals of different ages. However, visit ancient semi-natural woodland and typically you will find that one or two tree species predominate, either because environmental conditions favour them or because they have been encouraged by man. Although these woodlands do not qualify as habitats in the strict sense (ecologists refer to them as vegetation types), some of them are easily recognised and, more significantly, have a classic set of woodland animals, plants and fungi associated with them.

Professional ecologists recognise a large number of subtly different vegetation types and use them to define the natural history of the British and Irish landscape. However, for most people the differences between many of these types are too subtle to discern and most are happy to make reference to just a handful of easily recognised woodland types. The descriptions that follow cover the more distinctive, widespread and easily recognised of these communities.

MANAGED PEDUNCULATE OAK STANDARDS WITH HAZEL COPPICE

The Pedunculate Oak is one of the most common large native tree species across much of central and southern England, and in many woodlands it dominates in terms of stature and tree canopy cover. In some areas, its prevalence may be the result of soil conditions favouring its growth over other species. However, in many instances, an examination of the woodland structure and history will reveal man's contribution to the equation. If, for example, a particular woodland has oaks all of a similar size, then even if they are old their presence and dominance is likely to be the result of planting, in earlier centuries, and of selective felling of other species. This is hardly surprising because oak has always been considered the construction timber of choice. In such circumstances, mature oaks are referred to as *standards*.

In many semi-natural Pedunculate Oak woodlands you will find an understorey of Hazel. Typically this species will have been planted and managed on a regular basis by coppicing

A managed Pedunculate Oak woodland has near-continuous canopy cover. Only in winter and spring, before the leaves emerge, do significant light levels penetrate to the woodland floor.

(*see* p. 57) to produce a constant supply of straight branches, used in hurdle-making and other woodland crafts.

In spring, a sympathetically managed woodland of this type is a floral delight to the eye, with carpets of Bluebells, Wood Anemones and Wood-sorrel intermingling with patches of Dog's Mercury, Early-purple Orchids and Wild Daffodils. Pedunculate Oaks probably support the greatest diversity of invertebrate life of any of our trees, feeding beetles and moth larvae in abundance. If the occasional Sallow also grows among them, then there is a good

chance that Purple Emperor butterflies may be present, assuming the woodland lies within this species' restricted English range. The larvae feed on Sallow leaves but the adults – males in particular – depend on mature oak standards, using their canopies to define and defend territories. Birdlife is usually as rich as can be expected for a lowland woodland but arguably the star of the show is the Dormouse. Dependent on Hazel nuts in autumn, they occupy the oak canopy for much of spring and summer, feeding on oak flowers and insects.

ABOVE: A majestic male Purple Emperor surveys his domain from the canopy of a mature Pedunculate Oak.

RIGHT: Carpets of Bluebells are a stunning feature of managed Pedunculate Oak woodland in lowland Britain.

SESSILE OAK WOODLAND

Although Sessile Oak often grows alongside its Pedunculate cousin, its requirements for optimum growth are subtly different and it predominates in upland and western parts of the region where rainfall is highest. Often the best remaining tracts of Sessile Oak woodland are on steep slopes, in part because the trees are less easy to fell there; in such circumstances they are often referred to as 'hanging oak woodlands'. Try to negotiate the broken ground and slippery roots of one of these woodlands and it is not hard to see why.

Although the wildlife that Sessile Oak is capable of supporting is broadly similar to that of a Pedunculate Oak, in reality there are subtle differences. In part this may be a reflection of the generally harsher climatic conditions that it favours, although the history of land use by man sometimes plays a part too.

In lowland oak woods that have an understorey of Hazel, grazing animals were typically excluded in the past, or at the very least discouraged on account of the damage they might

RIGHT: Hanging Sessile Oak woodland.

BELOW: Sessile Oak woodland cloaking steep hillsides on Exmoor.

ABOVE: In the British Isles, Pied Flycatchers reach their highest breeding densities in Sessile Oak woodlands.

RIGHT: Tree Lungwort.

do to growing shoots. By contrast, many hanging woods of Sessile Oak are freely grazed by sheep and cattle with the consequence that the shrub layer is considerably reduced. Bluebells and other flowers of the woodland floor are common, but what strikes most visitors is often the abundance of ferns, mosses and liverworts. In part this is a result of an absence of competition from more palatable (in grazing animal terms) flowering plants, but the humidity of the regions where Sessile Oaks grow also plays a part. Hence the abundance of epiphytic mosses and liverworts in particular, often festooning trunks and branches well off the ground. Sessile Oak woodlands are the classic domain of nesting Pied Flycatchers and Redstarts.

ASH WOODLAND

Ash is a widespread tree in Britain, often growing alongside oaks and other deciduous species. However, in some circumstances it comes to dominate certain woodlands. It thrives if soil conditions suit it – it favours basic soils, and can tolerate occasionally waterlogged conditions, but will also grow on limestone pavements.

In lowland Britain in particular, Ash has often been deliberately encouraged and managed as a source of excellent timber. Traditionally, it was coppiced regularly to produce tall, straight poles and in some woods huge stools have developed that are hundreds of years old. Hazel coppice is often grown as an understorey beneath its larger cousin.

Ancient stools of Ash produce a succession of tall, straight poles if coppiced periodically and managed correctly.

Woodland flowers are very much a feature of traditionally managed Ash woods. Bluebell, Lesser Celandine, Dog's Mercury and Wood Anemone are often common, with star attractions being Herb-Paris, Goldilocks Buttercup and Early-purple Orchid.

Upland limestone pavement is perhaps a surprising place to find Ash, given its tolerance of, and seeming predilection for, damp ground in other locations. Nevertheless, thrive it does, often accompanied by Bird Cherry and Rowan.

Dark and gloomy during the summer months, by contrast a Beech woodland in winter is light and airy, with colour provided by the fallen leaves.

BEECH WOODLAND

Few trees are more successful at eliminating competition from rival species than Beech, and in mature woodland it often forms such a dense canopy that competition from other tree species is essentially eliminated. Beech will tolerate a wide range of soil types, from fairly acid to calcareous. It is arguably at its finest growing on the latter, especially when covering the slopes of chalk downs. Beech woodlands in such settings have the fitting name of 'hangers'.

During the summer months very little light penetrates to the woodland floor. Low light levels, combined with the dense carpet of fallen leaves that covers the ground, ensures that few woodland flowers are able to survive. Those that do are restricted mainly to clearings and rides, and in such places Sanicle, Woodruff and a number of helleborines can sometimes be found. A small group of specialist plants do grow in the deep shade of Beech woodlands: species such as the Bird's-nest Orchid and Yellow Bird's-nest lack chlorophyll and cannot photosynthesise, relying on a saprophytic way of life for nutrition.

For many people, Beech woods are at their best in the autumn, not only because of the stunning foliage colours but also because of the intriguing range and abundance (in wet years) of fungi. Many of these are entirely restricted to Beech woodlands and some of the associated *Russula* and *Boletus* species are extremely colourful themselves.

ABOVE: The Devil's Bolete is a rare and spectacular fungus that grows beneath ancient Beech trees on chalky soil.

RIGHT: Although it is seen only irregularly, the Ghost Orchid is the Holy Grail for British botanists. Scouring ancient Beech woods in central England represents a naturalist's only realistic prospect of finding one in this country.

BIRCH WOODLAND

Birches are often viewed with disdain and dismissed as being little more than scrub trees. It is true that, in certain circumstances, Silver Birch, in particular, is an aggressively invasive species of heathland and newly cleared woodland on neutral to acid soils. However, in its favour is the fact that it plays host to a wide range of insects, in particular the larvae of many moth species. Most feed on birch leaves but the larva of the Large Red-belted Clearwing *Synanthedon culiciformis* feeds on wood, sometimes being discovered in cut stumps.

In the autumn, Silver Birches usually turn a spectacular golden yellow for a week or two. Around the same time, an amazing array of fungi put in an appearance as well, many of them only found growing in association with these trees. The best-known fungal association with Silver Birch is probably the Fly Agaric *Amanita muscaria*, an unmistakable and striking species. But dozens of other species are invariably associated with these trees, including several *Boletus* and *Russula* species as well as the Birch Polypore *Piptoporus betulinus*.

ABOVE: Silver Birches provide stunning autumn colours that lift the spirits.

The larvae of the Large Red-belted Clearwing feed beneath birch bark and the secretive, day-flying adults are sometimes spotted resting among the foliage.

The Buff-tip has an uncanny resemblance to a snapped birch twig: a convincing camouflage that increases the moth's chances of avoiding detection by predators.

The Fly Agaric has a mycorrhizal association with the roots of birch trees (*see* p. 34), so you are unlikely to find this fungus growing anywhere else.

Downy Birch often grows alongside Silver Birch but comes to replace it in many western, northern and upland parts of the region. In terms of specifically associated wildlife, it has much in common with Silver Birch, particularly when it comes to fungi. However, in Scotland, in particular, superficially similar hardy Highland species often replace their southern counterparts. Look out for the intriguing Hoof Fungus *Fomes fomentarius* growing on stumps and trunks.

CARR WOODLAND

Nature is dynamic, and there is a general trend in all environments for newly created habitats to become colonised by increasingly more stable plant communities. In most locations this progression follows a predictable route and the process itself is referred to as *habitat succession.*

The impact of succession in wetland habitats is particularly striking. Marginal vegetation soon encroaches upon areas of open water, and silt builds up, gradually reducing water levels. Reeds, bulrushes, sedges and rushes then take hold and open water soon gives way to tussocky ground. The resulting mire is referred to as a fen if the water is neutral to basic or a bog if the ground is acidic. Trees soon begin to appear and those tolerant of waterlogged ground – Alder and various willow species – give rise to a woodland community called *carr* where once a fen prevailed. Carr is the penultimate stage in the process of succession, which leads eventually to the formation of dry land and a climax community of Oak and Ash.

Carr woodland is not the easiest of plant communities to explore and, unless the area in question is a boardwalked nature reserve, wet feet are inevitable. However, such areas are usually botanically rewarding. Insect life usually abounds too, particularly in the form of midges and mosquitoes.

The damp environment with which Alder carr is associated allows ferns and mosses to thrive in abundance.

NATIVE SCOTS PINE FOREST

Visit the Highlands of Scotland and you will discover remnants of the ancient Scots Pine forests that once cloaked the region; these areas are often referred to as Caledonian Pine forests. In contrast to the relative paucity of wildlife found in most regimented conifer plantations, there is plenty of wildlife to find here. In Britain, Crested Tits are found only in these forests and the Scottish Crossbill occurs nowhere else in the world. Pine Martens and Red Squirrels also occur in good numbers.

Native Caledonian Pine forests are typically rather open, comprising trees of varying ages, including many dead and dying individuals of course. Typically there is a lush ground cover of Bilberry and mosses, with botanical highlights that include several wintergreen species, Twinflower and Creeping Lady's-tresses, the latter a charming if diminutive orchid.

ABOVE: Native Scots Pine forests are typically light and airy, with an understorey of Juniper and Bilberry.

LEFT: Twinflower is a scarce but enchanting botanical highlight of a few Scottish forests.

ALIEN CONIFER FORESTS

If you encounter conifers in Britain outside the native range of the Scots Pine in central Scotland, then (with the exception of a very few Yew woods) their occurrence will not be natural. Regimented plantations are easy enough to spot, with trees of uniform age planted a standard distance from one another. But even isolated clumps of Scots Pine found on a southern heathland are at best going to be naturalised trees, their seeds having spread from nearby plantations.

Does the fact that most conifers in Britain are introduced matter? In the case of isolated trees or small clumps, perhaps not. But, when it comes to the impact upon native habitats and wildlife, then plantations are a different matter. The dense manner in which they are planted effectively destroys the natural vegetation that once existed there – just think of the scandalous destruction of areas of Scotland's Flow Country if you are in any doubt. But even naturalising conifers can be a threat, endangering already diminishing areas of heathland.

The benefits of conifer plantations to a select band of birds – Crossbills in particular – is often cited in favour of tolerance, although of course not all planted conifer species produce cones that are accessible to these birds. However, anyone willing to take an overall ecological view of the process, rather than trying simply to look on the bright side of things, will realise that the losses, in wildlife terms, heavily outweigh the gains.

FUNGI, TREES AND WOODLAND

Most naturalists recognise the significance of the role played by fungi in decay and the recycling of nutrients from dead organic matter back into the environment. But beyond that, for many, fungi are little more than colourful curiosities in autumn woodland. Worse still, people whose interest is primarily in trees often view fungi with suspicion, concerned about the threat of disease. True, a few fungal species are capable of destroying a precious specimen tree. However, the real relationship between trees and most fungal species (indeed between fungi and almost all plants) is entirely positive and profound, and illustrates the dangers of generalisation. It involves an intimately close, and mutually beneficial, relationship called *symbiosis*. That between birches and the Fly Agaric *Amanita muscaria* is often quoted, but in reality almost every trees species found in Britain depends for its survival upon a unique relationship between its roots and specific soil fungi.

Gardeners occasionally come across networks of fine threads in the soil. Essentially these are the body of a fungus, the *mycorrhiza*, a term that means 'fungus root'. Several different types of mycorrhizal relationship exist between fungi and plants but ectomycorrhizal ties are of particular significance to trees. In this relationship the fungus sheaths the tree roots and hyphae (fungal threads) extend into the soil and into the cortical layer of the root cells. The fungus derives almost all of its energy requirements, particularly as sugars, from the photosynthetic reactions of the tree's leaves. In return the tree obtains nitrogen and phosphorus, otherwise in short supply, via fungal action in the soil. Many of the toadstools that adorn the woodland floor in autumn are the fruiting bodies of these ectomycorrhizal fungi. This explains why, in many cases, certain toadstool species are found only under specific trees. For further information about the subject, and about fungi in general, see the excellent *Fungi* in the Collins New Naturalist series (*see* p. 312).

Honey Fungus may be dreaded in a garden context, but its role in the decomposition and recycling of nutrients from dead timber is profound.

In the case of a few fungal species, their relationship with trees is less benign and parasitism is well documented, as are instances where fungi invade an already diseased or damaged tree, hastening its end. It is a short step from this strategy to that found in the Honey Fungus *Armillaria mellea*, which is capable of penetrating seemingly healthy trees and ultimately causing their demise.

Arguably the most important environmental role played by fungi, of course, is in decomposition. Without the process of decay, nutrients locked up inside dead organic matter would remain there and the cycle of life would grind to a halt. Although fungi are not the only agents of decomposition (bacteria obviously have a significant part to play and the mechanical action of many insects is important too), their role is vital. In the context of trees and woodland, they ensure that fallen leaves are recycled and that nutrients contained within fallen timber enter the woodland soil.

Woodland fungi come in all shapes, sizes, colours and growing habits. Among the more intriguing are the Pinecone Cap *Strobilurus tenacellus* (A), which grows on decaying pine cones; the Sulphur Tuft *Hypholoma fasciculare* (B), which lives in decaying stumps; the Collared Earthstar *Geastrum triplex* (C), which appears in leaf litter, often under Beech or pines; the Blusher *Amanita rubescens* (D), which is associated with oaks; King Alfred's Cakes *Daldinia concentrica* (E), found on dead wood, often that of Ash; the Jelly Ear *Auricularia auricula-judae* (F), which grows on Elder branches; the Blue Roundhead *Stropharia caerulea* (G), found among fallen pine needles; and the Birch Polypore *Piptoporus betulinus* (H), which grows on birch trunks.

GALLS

People are often intrigued by the discovery of galls, peculiarly deformed growths found on many trees. Most are caused by gall wasps, a group of insects most of which are tiny and hence easily overlooked. They lay their eggs on tree buds, leaves or flowers (according to species), and galls form from plant tissue in response to chemicals produced by the gall wasp larvae, which live inside the gall. A gall's appearance is unique to each gall wasp species. Many species have two entirely separate generations, each one producing distinct galls.

Pedunculate Oak hosts a particularly large quota of gall wasp species, not to mention some striking galls. Among the more familiar are the following: Marble Galls, spherical and woody galls, up to 25mm across, caused by *Andricus kollari*; Oak Apples, spherical but knobbly galls, caused by *Biorhiza pallida* (the second generation form galls on oak roots); Artichoke Galls, where buds become swollen and artichoke-like, caused by *Andricus fecundator*; Spangle Galls, silk-button or flat button galls on the underside of leaves, caused by wasps of the genus *Neuroterus*; Cherry Galls, caused by members of the genus *Cynips*; Knopper Galls, distorted outgrowths on acorns, caused by *Andricus quercuscalicis*.

Although insects are the main causal agents of galls on trees in the British Isles, they are not alone in this ability. Many people will be familiar with the twiggy masses known as 'witches' brooms' that form on birch trees. These are caused by the fungus *Taphrina betulina*. A closely related species, *T. pruni*, causes Pocket Plum galls on Blackthorn, where ripening sloes become distorted to resemble miniature runner beans.

ABOVE: **Marble Galls.**

MIDDLE LEFT: **Oak Apples.**

MIDDLE RIGHT: **Artichoke Gall.**

FAR RIGHT: **Knopper Gall.**

LEFT: **Wasp of genus *Cynips*, causal host of Cherry Galls in oak.**

BELOW: **Cherry Galls.**

ABOVE: **Nail galls on Lime, caused by the mite *Eriophyes tiliae*.**

RIGHT: **Witches' brooms are most evident on leafless birches during the winter months. A large specimen can measure 40cm across or more.**

MOSSES, LIVERWORTS AND LICHENS

From a climatic perspective, trees help moderate temperature extremes and increase humidity within woodland. Unsurprisingly, this has a beneficial effect upon mosses, liverworts and lichens, many of which are prone to desiccation. Native woodland floors are often carpeted with these primitive plants, and fallen logs and rotting timber are soon cloaked in their fresh green growth.

In addition to their climatic effect, trees have the added benefit, ecologically, of creating a third dimension in any woodland. Bryophytes take full advantage of this added woodland complexity and, under certain circumstances, trunks and branches can be festooned with a lush growth of mosses, liverworts and lichens. When growing in such places, these plants are referred to as *epiphytes* – they grow on the surface of the tree, are not parasitic, and cause no damage to the host.

ABOVE: White Fork Moss *Leucobryum glaucum* forms sizeable clumps on undisturbed woodland floors, particularly under Beech.

LEFT: In western Britain, where annual rainfall is high, tree trunks are often completely cloaked in mosses and liverworts.

Visit almost any native woodland in Britain and you will find a profusion of mosses, liverworts and lichens growing both on the woodland floor and as epiphytes. However, for truly spectacular displays of epiphytic bryophytes you will need to visit woodlands in the west of the country, particularly those that grow on west-facing, rain-soaked hills or on the coast. In places it can be almost impossible to discern bare bark, with trunks, branches and twigs simply dripping with epiphytes. They cause no harm to the trees, other than the added weight perhaps, and gain their nutrition from nutrient run-off and from rain. Of course, the same is true of that other familiar epiphyte Ivy (a flowering plant), which is often maligned but hugely important in wildlife terms.

HEDGEROWS

In a few cases, hedgerows that we see today are the linear remnants of ancient forests, all that is left of woodlands cleared for early agriculture from the Iron Age onwards. However, most hedgerows have been deliberately planted by man and are often hundreds of years old. For many people, they form the visual backbone of the British and Irish countryside.

Some hedgerows were planted to mark parish boundaries and most of these were in existence by the time of the Norman Conquest. However, many more were planted during the process of land enclosure, mainly between the 16th and 18th centuries. Before this time, much of the landscape was open land, used mainly for rough grazing with a little piece-meal agriculture. Enclosure, land clearance and hedge planting helped define and demar-cate land ownership and in many cases hedges were maintained as stockproof barriers.

Ancient hedgerow.

When first created, hedgerows would probably have been planted with easily available, locally sourced shrubs and trees noted for their branching habit and, in some cases, spines. No surprise, then, that we find shrubs such as Blackthorn, Hawthorn and willow species predominating. Various elm species were popular hedgerow shrubs and trees in many parts, particularly East Anglia, while Beech prevailed in parts of the West Country, notably on Exmoor.

Over time, of course, natural additions to most hedgerows have occurred, seeds being transported by the wind or by roosting birds, depending on the method of seed dispersal of the tree or shrub in question. Given time, a hedgerow takes on the species composition of neighbouring woodland and much of the wildlife associated with woodland edges too. Research has found that there is a reasonably well-defined correlation between the number of woody species in a hedgerow and its age. The exact formula for calculating a hedgerow's age is complicated, but a rough guide gives a reasonable degree of accuracy: count a 30m stretch of hedgerow and if 10 woody species are present, then it is likely to be around 1,000 years old; each woody species represents approximately 100 years.

Most hedgerows are man-made, and people control their fate to this day. Regular man-agement – proper laying rather than indiscriminate flailing – is required to keep a hedge in good order. But, sadly, that seldom happens and, worse still, vast lengths of hedgerows have already disappeared, grubbed up by farmers in the mid-20th century; they were seen as a threat to farming 'efficiency'. Today, grants for hedgerow planting schemes are, to some extent, redressing the balance.

The ecological importance of hedgerows is hinted at by the number of native plant and animal species with 'hedge' in their names: Hedge-sparrow (an old name for the Dunnock), Upright Hedge-parsley, Hedge Bindweed, Hedge Bedstraw, Hedge Brown (an old name for the Gatekeeper butterfly) spring to mind. A dense hedgerow is an ideal location for nesting birds such as Yellowhammers, Chaffinches and Goldfinches, while innumerable

Not only does a badly flailed hedge (top) look unsightly, but its value to wildlife is degraded and frequently it ceases to be stockproof as well. By contrast, within a couple of years, a well-laid hedge (above) forms an impenetrable, wildlife-rich barrier that lasts for a decade or more without the need for further work.

insects also make it their home, the species present dependent upon the exact composition of the hedge in question. And, of course, the herb layer associated with most hedgerows is also important, both for the plants that it comprises and for the invertebrates associated with it.

WINTER TWIGS

If you have the time and the patience, identifying native deciduous trees and shrubs is not too daunting during periods of the year when there are leaves and flowers or fruits to help. But during the winter months, identification is considerably more challenging. Challenging, but not impossible, because the texture and colour of the twigs themselves and their buds, as well as the way in which they branch, all give clues to a tree's identity.

Identifying winter twigs is not just an academic exercise. It can have practical applications too. Imagine, for example, that you want to cut back a native hedge (best done in the winter) but do not want to damage certain key tree and shrub species. Or you might want to undertake some woodland conservation work to remove alien species but leave notable native trees or shrubs untouched. Unless you had the foresight to identify and mark every tree and shrub back in the summer months, you will need to be able to recognise the trees and shrubs by their winter twigs alone. The following pages illustrate some of the more common and widespread native and widely naturalised trees and shrubs. For ease of reference, I have divided them into three sections: classic hedgerow shrubs and trees; woodland trees; and garden and park trees that are sometimes naturalised. Of course, these are artificial divisions so be aware, for example, that some hedgerow shrubs and trees may be found in woodland and vice versa.

CLASSIC HEDGEROW SHRUBS AND TREES

OPPOSITE BUDS

ELDER *Sambucus nigra*
Curved, greyish and warty twigs with purplish buds.

GUELDER-ROSE *Viburnum opulus*
Straight, hairless and greyish, angled in cross section; buds scaly, reddish and paired. Terminal fruit stalk often persists.

WAYFARING-TREE *Viburnum lantana*
Straight, downy and yellowish brown, rounded in cross section; buds non-scaly, yellowish and paired.

SPINDLE *Euonymus europaeus*
Straight, stiff and greenish brown, slightly angled in cross section; buds paired, ovoid to conical and greenish brown.

DOGWOOD *Cornus sanguinea*
Straight, stiff and reddish, downy at first but then shiny; buds reddish and flattened-conical at first.

BUCKTHORN *Rhamnus cathartica*
Straight and greyish to yellowish brown; buds conical to talon-like, reddish brown and in slightly staggered, opposite pairs.

BLACKTHORN *Prunus spinosa*
Straight and stiff, purplish and shiny but often coated with a bloom of green algae; side shoots terminate in sharp spines. Buds small, reddish and ovoid.

WILD PRIVET *Ligustrum vulgare*
Straight and greyish to yellowish brown,
and slightly rough; buds reddish green,
ovoid and borne in staggered, opposite pairs.

FIELD MAPLE *Acer campestre*
Straight to slightly curved, greyish to yellowish
brown; buds paired, yellowish brown with
greyish hairs.

ALTERNATE BUDS

COMMON HAWTHORN *Crataegus monogyna*
Stiff, reddish brown or greenish with 1–2cm
spines and knobbly brown buds.

MIDLAND HAWTHORN *Crataegus laevigata*
Similar to Common Hawthorn but twigs less
stiff and less spiny (often spineless).

HAZEL *Corylus avellana*
Mainly straight (but sometimes zigzag towards
tip, between nodes), greenish brown and
sparsely hairy; buds green and ovoid. Male
catkins usually present by mid-January; tiny
reddish female flowers sessile.

ALDER BUCKTHORN *Frangula alnus*
Straight, with side branches widely spreading,
purplish brown with white streaks created
by elongated lenticels; buds alternate and
scale-less, with tufts of orange hairs.

WEEPING WILLOW *Salix × sepulcralis*
Straight and pendulous, greyish at first but
yellowish later (golden in some cultivars); buds
yellowish and narrow-conical to talon-like.

CRACK WILLOW *Salix fragilis*
Straight, yellowish brown or reddish, and
downy at first but soon hairless and shiny;
buds narrow, flattened and smooth.

GOAT WILLOW *Salix caprea*
Greyish to yellowish brown with a rough
texture at first, but smooth later; buds
yellowish, ovoid to clog-like, and rather
congested towards the shoot tip.

WHITE WILLOW *Salix alba*
Slender, straight and greyish, downy at first
but smooth later; buds flattened and silky.

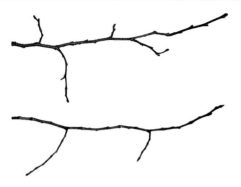

ENGLISH ELM *Ulmus procera*
Slender and downy; often zigzag at the nodes.
Buds tiny, and ovoid to spherical.

WYCH ELM *Ulmus glabra*
Dark greyish brown with coarse hairs; twigs
often look knobbly and are not straight, with
side shoots often reflexed. Buds tiny and dark
brown.

WOODLAND TREES

OPPOSITE BUDS

SYCAMORE *Acer pseudoplatanus*
Reddish brown, often tinged greenish,
marked with lenticels; buds ovoid, swollen
and greenish.

NORWAY MAPLE *Acer platanoides*
Yellowish grey to pinkish brown, smooth
with lenticels; buds ovoid and reddish.

HORSE-CHESTNUT *Aesculus hippocastanum*
Thick and straight, reddish grey and marked
with horseshoe-shaped leaf scars; buds
reddish brown, pointed and sticky.

ASH *Fraxinus excelsior*
Stout and greyish, swollen below buds, and
sometimes yellowish there; buds blackish
and mitre-shaped.

ALTERNATE BUDS

ASPEN *Populus tremula*
Straight between nodes, but overall rather
arching, yellowish brown and shiny eventually;
buds long and sharply pointed.

BLACK-POPLAR *Populus nigra* ssp. *betulifolia*
Knobbly and irregular, greyish in older
sections of twigs but new growth yellowish;
buds yellowish brown and sharply pointed.

GREY POPLAR *Populus × canescens*
Rather irregular, yellowish grey, downy at first
but soon smooth and shiny; buds yellowish
brown, narrow and very sharply pointed.

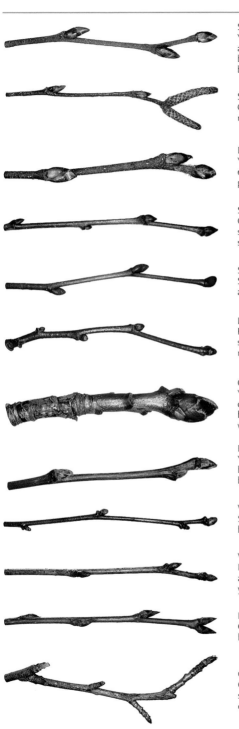

SILVER BIRCH *Betula pendula*
Yellowish to purplish brown and hairless, although white warts create a rough texture; buds ovoid and pointed, greenish and reddish brown.

SILVER BIRCH *Betula pendula*
Catkins appear on bare twigs by late winter; male and female flowers in separate catkins.

DOWNY BIRCH *Betula pubescens*
Yellowish brown to purplish, with downy hairs especially towards shoot tips; buds ovoid and greenish and reddish brown.

SWEET CHESTNUT *Castanea sativa*
Greyish yellow with a few, elongated lenticels, swollen below the buds; buds yellowish red, swollen and pointed.

SMALL-LEAVED LIME *Tilia cordata*
Straight between nodes but zigzagging, shiny and reddish; buds swollen and reddish.

LIME *Tilia × europaea*
Reddish to yellowish brown, hairless and shiny; buds ovoid to spherical, swollen and reddish.

COMMON WHITEBEAM *Sorbus aria*
Yellowish to reddish, or greyish in shade, downy at first but soon hairless and shiny; buds ovoid, the reddish-brown scales tipped with greyish hairs.

ROWAN *Sorbus aucuparia*
Yellowish brown to purplish, downy at first but soon smooth; buds conical and purplish brown but scales have long, greyish hairs.

WILD SERVICE-TREE *Sorbus torminalis*
Slender, shiny and yellowish to greyish brown; buds green and spherical to ovoid.

WILD CHERRY *Prunus avium*
Rather straight, yellowish brown but with a grey bloom; buds narrow-conical and yellowish to purplish brown.

BIRD CHERRY *Prunus padus*
Greyish brown with paler, elongated lenticels; buds narrow, elongated and pointed.

CRAB APPLE *Malus sylvestris*
Greyish brown or yellowish, hairless and shiny; buds ovoid and reddish brown, with downy tips.

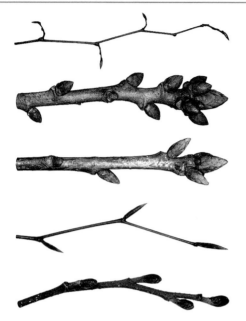

HORNBEAM *Carpinus betulus*
Slender and irregular, hairy at first but soon smooth; buds slender and pointed.

PEDUNCULATE OAK *Quercus robur*
Greyish to pale purplish and often rather knobbly; buds orange-brown and ovoid.

SESSILE OAK *Quercus petraea*
Similar to Pedunculate Oak but more slender; greyish to pale purplish and often rather knobbly; buds buffish brown, ovoid and rather scaly.

BEECH *Fagus sylvatica*
Slender and greyish brown, zigzagging at nodes; buds extremely slender and pointed, and orange-brown.

COMMON ALDER *Alnus glutinosa*
Curved and yellowish grey, hairless with pale lenticels; buds purplish brown, mealy, ovoid and stalked, the result looking rather club-like.

GARDEN AND PARK TREES THAT ARE SOMETIMES NATURALISED

ALTERNATE BUDS

CULTIVATED APPLE *Malus domestica*
Greyish, slightly hairy and often rather knobbly, irregular and gnarled-looking; buds pointed, with greyish woolly hairs.

COMMON PEAR *Pyrus communis*
Rather straight and shiny reddish brown; usually spineless (Wild Pear similar but spiny). Buds tiny, brown and pointed.

PLUM *Prunus domestica* ssp. *domestica*
Rather straight and reddish purple, downy at first but soon smooth; buds small, ovoid and clustered.

CHERRY PLUM *Prunus cerasifera*
Rather straight, greenish and hairless; buds tiny, ovoid and clustered.

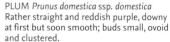

LONDON PLANE *Platanus × hispanica*
Greenish and usually slightly curved; buds alternate, conical, smooth and green, tinged reddish.

COMPARING LEAVES OF COMMON TREES AND SHRUBS

Although somewhat variable, leaves are probably the most useful features for identifying a tree. Regardless of age or growing position, the leaves of most species have a set of characteristics unique enough to allow separation from even closely related trees. The following leaves from common native and widely naturalised trees and shrubs will allow naturalists to put a name to most species found growing in their local hedgerows and woodlands.

SIMPLE LEAVES, WITH ENTIRE (OR MINUTELY TOOTHED) MARGINS

BEECH *Fagus sylvatica*
Oval and pointed, to 10cm long, with a wavy margin and a fringe of silky hairs when freshly open. Fresh green when newly emerged, dark green in summer, orange-brown in autumn.

ALDER BUCKTHORN *Frangula alnus*
Broadly oval, to 7cm long, with up to 9 pairs of veins, curving towards the margin. Glossy green above and paler below, turning lemon-yellow in autumn, or redder if exposed to bright sunlight.

ALDER *Alnus glutinosa*
Stalked and noticeably rounded, to 10cm long, usually with a notched apex and a wavy margin; 5–8 pairs of veins with long hairs in axils on underside of leaf.

DOGWOOD *Cornus sanguinea*
Oval and pointed, with 3–4 pairs of prominent veins. If a leaf is snapped and the 2 halves are gently pulled apart, stringy latex appears where veins were broken and connects 2 halves of leaf. Leaves become a rich, deep red in autumn.

WILD PRIVET *Ligustrum vulgare*
Shiny, oval and opposite, up to 6cm long. Semi-deciduous.

COMMON PEAR *Pyrus communis*
Oval to elliptical, to 8cm long; margins with minute teeth, and leaves smooth and almost glossy when mature.

SIMPLE LEAVES, WITH TOOTHED MARGINS

BLACKTHORN *Prunus spinosa*
Oval and pointed at tip, to 4.5cm long, petiole 1cm long; upper surface smooth and dull green, lower surface downy along the prominent veins.

BUCKTHORN *Rhamnus cathartica*
Ovate or nearly rounded with a short pointed tip, to 6cm long and 4cm wide, finely toothed around margin, glossy green above and pale below. Conspicuous veins on upper surface converging towards tip of leaf. In autumn leaves turn yellow.

SPINDLE *Euonymus europaeus*
Ovate, to 10cm long, with a pointed tip and sharply toothed margins. A rich shade of purple-orange in autumn.

WILD CRAB *Malus sylvestris*
Ovate to obovate, up to 11cm long, smooth above and below when fully open.

CULTIVATED APPLE *Malus domestica*
Elliptical and rounded at the base with a slightly pointed tip, to 13cm long; slightly downy above and normally very downy below.

WILD CHERRY *Prunus avium*
Ovate with a long pointed apex and forward-pointing irregular teeth on margins; upper surface smooth and dull, lower surface downy on veins. Petiole 2–5cm long, with 2 glands near leaf junction.

BIRD CHERRY *Prunus padus*
Elliptical to elongate, to 10cm long, finely toothed on margins and tapering at tip; tough, dark green above and slightly blue-green below.

CHERRY PLUM *Prunus cerasifera*
Ovate, tapering at base and tip, to 7cm long; margins with numerous rounded teeth and underside with downy veins. Petiole 1cm long, pinkish, grooved.

CHERRY PLUM *Prunus cerasifera* var. *atropurpurea*
Identical to ordinary form of Cherry Plum except that leaves are maroon from spring onwards.

PLUM *Prunus domestica* ssp. *domestica*
Ovate, up to 8cm long, with toothed margins, a smooth green upper surface and a downy lower surface.

SILVER BIRCH *Betula pendula*
Triangular and pointed, to 7cm long, with large teeth separated by many smaller teeth; thin and smooth when mature. Petiole hairless. Leaves turn golden yellow in autumn.

DOWNY BIRCH *Betula pubescens*
More rounded at base than those of Silver Birch and more evenly toothed; white hairs in axils of veins on underside, and petiole hairy.

SWEET CHESTNUT *Castanea sativa*
Glossy, lanceolate, to 25cm long, margins serrated with spine-tipped teeth, pointed at tip, and sometimes with a slightly heart-shaped base.

HORNBEAM *Carpinus betulus*
Oval and pointed with a rounded base, short petiole, and double-toothed margin; 15 pairs of veins, hairy below. Leaves turning yellow, through orange to russet-brown in autumn.

WHITEBEAM *Sorbus aria*
Oval, to 12cm long, and very hairy, especially on white underside; 10–14 pairs of veins.

WAYFARING-TREE *Viburnum lantana*
Ovate, to 14cm long, rough to touch; undersides thickly hairy with more stellate hairs.

SMALL-LEAVED ELM *Ulmus minor* ssp. *minor*
Superficially Hornbeam-like, leathery, to 15cm long, oval, pointed at tip, with toothed margins; unequal leaf bases, narrowly tapering on short side, and a short petiole.

ENGLISH ELM *Ulmus procera*
Rounded or slightly oval with short tapering tip; base unequal, longest side not reaching beyond petiole to twig. Leaf rough to touch; petiole (1–5mm long) and midrib finely downy.

WYCH ELM *Ulmus glabra*
Rounded or oval, to 18cm long, with a long tapering point at tip. Base of leaf unequal; long side of leaf base extends beyond the petiole (which is 2–5mm long) to twig. Leaves feel rough; upper surface hairy, and lower surface with softer, sparser hairs.

HAZEL *Corylus avellana*
Rounded, to 10cm long, with a heart-shaped base and pointed tip; margins double-toothed and upper surface hairy. Undersides of leaf veins with white hairs. Petiole short and hairy, and whole leaf has a bristly, rough feel.

GOAT WILLOW *Salix caprea*
Oval, to 12cm long, with a short twisted point at tip; dull green and slightly hairy above, grey and woolly below. Leaf margins have small, irregular teeth, and short petiole sometimes has 2 ear-like sinuous stipules at its base.

CRACK WILLOW *Salix fragilis*
Long, very narrow and glossy, with toothed margins; lower surface less glossy and slightly paler than upper surface. Petiole short and green.

WHITE WILLOW *Salix alba*
Long and very narrow, smaller than similar Crack Willow, bluish grey and silky-hairy above at first.

SMALL-LEAVED LIME *Tilia cordata*
Rounded, with a pointed tip, heart-shaped base and finely toothed margin, to 9cm long; dark shiny green and smooth above, paler and smooth below but with tufts of darker hairs in vein axils. Petiole smooth and up to 4cm long.

LIME *Tilia × europaea*
Broadly ovate with a short pointed tip, heart-shaped base and toothed margin, to 10cm long. Dull green above and paler below with tufts of white hairs in vein axils.

ASPEN *Populus tremula*
Rounded to slightly oval, with shallow marginal teeth. Green on both surfaces, but paler below, on long, flattened petioles. In autumn, leaves may turn golden yellow.

BLACK-POPLAR *Populus nigra* ssp. *betulifolia*
Triangular and long-stalked with a finely toothed margin; fresh shiny green on both surfaces.

GREY POPLAR *Populus × canescens*
Rounded to oval with regular blunt, forward-pointing teeth; borne on long petioles. Glossy grey-green above, lower surface with a greyish-white felt.

HOLLY *Ilex aquifolium*
Tough and leathery, to 12cm long, waxy above and paler below; margins variably wavy and spiny; leaves from upper branches of a large tree often flat and mostly spineless.

LOBED AND PALMATE LEAVES

SESSILE OAK *Quercus petraea*
Lobed, flattened, dark green and hairless above, paler below with hairs along veins; on yellow stalks, 1–2.5cm long, and lacking auricles at the base, distinguishing them from those of Pedunculate Oak.

PEDUNCULATE OAK *Quercus robur*
Deeply lobed, dark green and hairless above, with 2 auricles at the base; on very short stalks (5mm or less).

COMMON HAWTHORN *Crataegus monogyna*
Roughly ovate and deeply lobed, to 4.5cm long, usually with 3 segments; lobes pointed with just a few teeth near apex. Leaves feel tough. Dark green above, paler below, with a few tufts of hairs at axils of veins. Petiole about 2cm long and tinged pink.

MIDLAND HAWTHORN *Crataegus laevigata*
Superficially similar to Common Hawthorn, to 6cm long, but not as deeply or conspicuously lobed; lobes more rounded and toothed to the base.

GUELDER-ROSE *Viburnum opulus*
To 8cm long, with 3–5 irregularly toothed lobes and thread-like stipules. Leaves often turn a deep wine-red in autumn.

FIELD MAPLE *Acer campestre*
Usually strongly 3-lobed, to 12cm long; lobes themselves often have lobed margins and tufts of hair in axils of veins on underside. Newly opened leaves pinkish, becoming dark green and rather leathery later and bright yellow in autumn.

NORWAY MAPLE *Acer platanoides*
Bright green, smooth, to 15cm long with 5–7 toothed and sharply pointed lobes; lowest pair of lobes smaller than others. Note white hairs in axils of veins on paler underside of leaf.

SYCAMORE *Acer pseudoplatanus*
To 15cm long, and divided into 5 toothed lobes. Immature and fast-growing trees have deeply cut leaves and long scarlet petioles, whereas older trees have smaller leaves with shallower lobes and shorter pink or green petioles.

WILD SERVICE-TREE *Sorbus torminalis*
To 10cm long with 3–5 pairs of pointed lobes and a sharply toothed margin; basal lobes projecting at right angles, other lobes pointing forwards. Leaves turn to shades of red and russet in autumn.

LONDON PLANE *Platanus × hispanica*
To 24cm long and mostly 5-lobed and palmate; very variable, however, and degree of lobing may differ greatly.

HORSE-CHESTNUT *Aesculus hippocastanum*
Large, long-stalked and palmate, with up to 7 leaflets, each to 25cm long (central leaflets longest), sharply toothed and elongate-oval; upper surface mostly smooth, lower surface slightly downy.

COMPOUND LEAVES

ROWAN *Sorbus aucuparia*
Pinnate, composed of 5–8 pairs of toothed leaflets, each one to 6cm long, ovoid and markedly toothed; central rachis rounded near base, and grooved between leaflets.

ELDER *Sambucus nigra*
Usually with 5–7 (occasionally 9) pairs of leaflets, each one to 12cm long, ovate and pointed with a sharply toothed margin and slightly hairy underside; green in summer but sometimes turning deep plum-red before falling in autumn

ASH *Fraxinus excelsior*
Pinnate, to 35cm long with a flattened central rachis, which may be hairy, bearing 7–13 ovate-lanceolate, pointed and toothed leaflets, each one up to 12cm long; upper surface usually dark green and lower surface paler with densely hairy midribs.

NEEDLES AND NEEDLE-LIKE LEAVES

COMMON YEW *Taxus baccata*
Flattened, needle-like and up to 4cm long and 3mm wide, narrowing to a sharp point. Dark glossy green above and paler below with 2 pale yellowish bands. Leaves arising spirally around twig but flattened to lie in a row on either side of twig.

DOUGLAS-FIR *Pseudotsuga menziesii*
Needles, to 3.5cm long, blunt or slightly pointed, dark green and grooved above, with 2 white bands below.

LAWSON'S CYPRESS *Chamaecyparis lawsoniana*
Small and scale-like, up to 2mm long and flattened along shoot, in opposite pairs, and showing paler colours on underside of shoot. Crushed leaves smell of parsley.

LEYLAND CYPRESS × *Cupressocyparis leylandii*
Pointed, scale-like and about 2mm long.

SCOTS PINE *Pinus sylvestris*
Needles, in bunches of 2, grey-green or blue-green, up to 7cm long, usually twisted with a short point at the tip.

CORSICAN PINE *Pinus nigra* ssp. *maritima*
Soft, narrow needles, pale green, to 15cm long, often twisted in young trees.

COMPARING THE BARK OF COMMON TREES AND SHRUBS

At first glance, the bark of one tree can look pretty much like that of another. But, as with most things botanical, the more you look the more you will begin to discern similarities between the barks of trees of the same species, and differences between unrelated ones.

At this point it is worth raising a note of caution. The wise tree enthusiast will accept the fact that some individual trees cannot be identified by bark alone, for a number of reasons. Species such as Ash have bark that looks strikingly different depending on the growing conditions of the tree in question. Furthermore, in many species, the bark of young trees differs from older individuals, usually in being smoother and less rugged. It is also important to remember that the bark chemistry of certain tree species suits the needs of lichens and mosses, some of which are extremely colourful and can mask the true character of the bark. To check true bark colour, it is often necessary to scrutinise the shadiest side of a trunk.

With the above reservations, reasonably distinctive bark characteristics usually allow most common and widespread trees and certain larger shrubs to be identified with certainty. This ability is particularly valuable for people wishing to undertake conservation management of woodlands in winter and, with this in mind, the following images cover the commonest native and widely naturalised species to be found in Britain.

TREES

PEDUNCULATE OAK *Quercus robur*
Grey, with shallow ridges in young trees (A), becoming thick and deeply fissured with knobbly ridges in mature trees (B) and gnarled and rugged in ancient trees (C). The white spots in (B) are lichens; the green and blue patches in (C) are lichens and algae.

ASH *Fraxinus excelsior*
In young trees (A) smooth and pale grey (seen here with whitish lichens and yellowish algae), but in older trees (B) it becomes vertically fissured, although remaining grey-brown; the true colour is often obscured by the large colonies of lichens and mosses that grow well on mature Ashes in unpolluted areas (C).

SESSILE OAK *Quercus petraea*
Greyish brown with shallow reddish fissures when young (A). Most trees acquire deep vertical fissures and knobbly ridges with age, the bark coloured by algae, as here (B), but some have rather shallow scales.

SILVER BIRCH *Betula pendula*
Orange-red in young trees (A). In old trees, thick bark becomes deeply fissured at the base of the bole, often developing a pattern of dark diamond shapes (B); higher up the bole the bark is a smooth silvery white (C), often flaking away and revealing greyer patches below.

DOWNY BIRCH *Betula pubescens*
Purplish brown when young (A), becoming mostly smooth and brown or greyish with age and not breaking up into rectangular plates at the base like Silver Birch. The pale blue-grey patches in this picture (B) are lichens, the greenish epiphytes are mosses.

COMMON ALDER
Alnus glutinosa
Brownish and fissured into square or oblong plates that are of reasonably even width; the pale blue-grey patches in this picture are lichens.

BEECH *Fagus sylvatica*
Smooth and grey in young trees (A), often with horizontal lines, but becoming rougher with shallow ridges in older specimens (B).

WYCH ELM *Ulmus glabra*
Smooth and greyish in younger trees (A), becoming browner with deep, mostly vertical cracks and ridges with age (B).

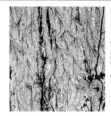
DUTCH ELM
Ulmus × hollandica
Brown, cracking into small, shallow plates; the greenish coloration of the bark in this picture is caused by lichens.

BIRD CHERRY
Prunus padus
Smooth dark and grey-brown, never peeling or fissuring. The greenish coloration in this picture is caused by algae and mosses. The bark releases a strong, unpleasant smell if rubbed.

WILD CHERRY
Prunus avium
Reddish brown and shiny, with circular lines of lenticels; in older trees, peeling horizontally into tough papery strips, and occasionally becoming fissured. The greenish coloration seen here is caused by lichens.

WILD CRAB
Malus sylvestris
Deep purplish brown, and cracking into small oblong plates; the green patches in this picture are mosses and algae.

SWEET CHESTNUT *Castanea sativa*
Silvery and smooth in young trees with fine vertical fissures (A); in older trees becoming more deeply fissured with grooves becoming markedly spiralled up the trunk (B).

HORSE-CHESTNUT
Aesculus hippocastanum
Smooth and pinkish grey in young trees but soon greyish brown, often flaking away in large, rather rounded scales.

BLACK-POPLAR
Populus nigra ssp. *betulifolia*
Deeply fissured and usually grey-brown (much darker in some old trees). The orange patches in this picture are lichens.

ASPEN *Populus tremula*
Smooth and greyish, with small, dark diamonds at first (A), becoming brown, ridged and fissured with age (B).

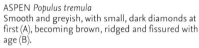

WILD SERVICE-TREE
Sorbus torminalis
Greyish brown at first, becoming darker with age and finely fissured into squarish to oblong brown plates.

ROWAN *Sorbus aucuparia*
Silvery grey with horizontal lenticels; usually smooth but sometimes feeling slightly ridged. The pale patches in this picture are lichens and the yellowish coloration is algae.

WHITE POPLAR
Populus alba
Pale in young specimens and broken by diamond-shaped scars; in mature trees deeply fissured on lower part of trunk. The greenish coloration in this picture is caused by algae, lichens and mosses.

GREY POPLAR *Populus × canescens*
Similar to White Poplar. Whitish with diamond-shaped fissures in young trees (A), but darker grey-brown and deeply fissured at the base of older trees (B).

COMMON WHITEBEAM *Sorbus aria*
Smooth and purplish grey in young specimens (A) but becoming ridged and fissured in older specimens (B).

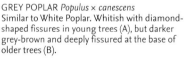

LONDON PLANE
Platanus × hispanica
Greyish brown and thin, flaking away in rounded patches to leave paler, yellowish areas beneath.

LIME *Tilia × europaea*
Grey-brown, becoming knobbly and ridged with age; lower trunk often shrouded by suckers and shoots and difficult to observe.

FIELD MAPLE
Acer campestre
Grey-brown and fissured with a slightly corky texture. The green coloration in this picture is caused by algae.

COMMON PEAR
Pyrus communis
Greyish brown, breaking up into small square plates that may fall away to reveal darker brown areas below.

SYCAMORE
Acer pseudoplatanus
Greyish and broken up by numerous fissures into irregular patches that sometimes fall away, leaving more orange-coloured areas beneath.

CULTIVATED APPLE
Malus domestica
Usually grey and brown but fissured and broken into scaly plates in older trees. Bark often coated in lichens, as in this picture.

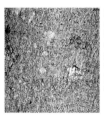

NORWAY MAPLE
Acer platanoides
Smooth and pinkish grey with regular but rather shallow ridges. The whitish and bluish patches in this picture are lichens.

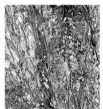

PLUM *Prunus domestica*
ssp. *domestica*
Dull brown, sometimes tinged purple, with deep, irregular and often swirly fissures developing with age. The pale patches in this picture are lichens.

HORNBEAM
Carpinus betulus
Silvery grey with faint orange or yellowish vertical lines; deeply fissured only towards the base.

HOLLY *Ilex aquifolium*
Smooth silvery grey to pinkish grey with fissures and tubercles appearing with age. The yellowish coloration seen in this picture is caused by algae.

SMALL-LEAVED LIME
Tilia cordata
Smooth and grey on young trees, but becoming darker purplish grey and cracked in older trees, often breaking away in flakes.

YEW *Taxus baccata*
Rather smooth and variably reddish, purplish and grey, sometimes peeling to reveal reddish-brown patches beneath. Shoots often appear on even large trunks.

CORSICAN PINE
Pinus nigra ssp. *maritima*
Greyish brown to purplish grey, rough in older trees with scaly plates and wide fissures. Austrian Pine *P. n.* ssp. *nigra* very similar but with darker bark.

DOUGLAS FIR
Pseudotsuga menziesii
Smooth and greyish green at first, with resinous blisters in young trees; developing orange fissures with age.

LAWSON'S CYPRESS
Chamaecyparis lawsoniana
Reddish to purplish brown, cracking vertically into long greyish plates in older trees.

SCOTS PINE
Pinus sylvestris
Reddish or greyish brown low down on the trunk, but markedly red or orange higher up in mature trees. The lower trunk is scaly, becoming more papery higher up.

EUROPEAN LARCH
Larix decidua
Rough and greyish brown in young trees, becoming pinkish brown and fissured with age, with scaly ridges. Japanese Larch *L. kaempferi* similar but with shaggier ridges.

SMALLER TREES AND SHRUBS

GOAT WILLOW
Salix caprea
Greyish, with diamond-shaped marks in young trees but soon deeply ridged, especially towards the base. Yellowish patches in this picture are caused by algae; blue-grey patches are lichens.

WHITE WILLOW
Salix alba
Deeply and rather irregularly fissured; dark greyish brown although often appearing paler (as in this picture) because of surface algae and lichens.

CRACK-WILLOW *Salix fragilis*
Dull grey-brown covered with thick interlocking criss-crossed ridges, even when fairly young (A); becoming extremely rugged and stained purplish and yellowish with age (B).

HAZEL *Corylus avellana*
Smooth and often shiny, peeling horizontally into thin papery strips. A coating of algae has turned the bark in this picture yellowish and matt.

BLACKTHORN
Prunus spinosa
Dark purplish brown but larger stems often looking paler because of a coating of algae and lichens, as in this picture.

CHERRY PLUM
Prunus cerasifera
Dark purplish grey (sometimes coated with greenish algae, as in this picture), becoming fissured; peeling with age.

ALDER BUCKTHORN *Frangula alnus*
Young stems (A) purplish grey with pale, knobbly areas; in older trees (B) smooth, grey and vertically furrowed, but with colour often obscured by algae and lichens, as in this picture.

BUCKTHORN
Rhamnus cathartica
Dark orange-brown, becoming almost black in older trees, but still revealing orange patches between the numerous fissures. True colour often obscured by lichens, as in this picture.

DOGWOOD
Cornus sanguinea
Grey and smooth, with snakeskin-like pattern of reddish ridges. In this picture the colour is obscured by algae; the pale patches are lichens.

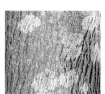

SPINDLE
Euonymus europaeus
Smooth and grey, becoming slightly fissured and tinged pink as the tree ages. Bark colour often obscured, as here, by algae and lichens.

HAWTHORN
Crataegus monogyna
Usually heavily fissured into a fairly regular pattern of vertical grooves, the bark sometimes cracking into plates; outer layers greyish brown, lower layers more orange.

MIDLAND HAWTHORN
Crataegus laevigata
Greyish to yellowish brown, rather smooth in young stems (as here) but cracking into regular-shaped plates with age, revealing darker, browner patches beneath.

GUELDER-ROSE
Viburnum opulus
Buffish brown, sometimes tinged red, with an intricate network of shallow ridges. The greenish areas in this picture are mosses and algae.

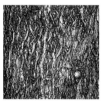

WAYFARING-TREE
Viburnum lantana
Bluish grey with an attractive pattern of orange patches in young stems (shown here); becoming browner and more uniform with age.

ELDER *Sambucus nigra*
Greyish brown (A) and becoming deeply grooved and furrowed with age; often takes on a corky texture in older specimens (B). Often heavily coated with orange-yellow lichens, making whole tree conspicuous in winter.

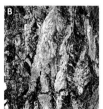

BOX
Buxus sempervirens
Smooth and grey, breaking into small squares with age; edges of these squares rounded off, not sharp.

WILD PRIVET
Ligustrum vulgare
Reddish brown (but often coated with algae, as in this picture), with striking indentations (revealing orange layers) that look as though they were caused by an axe.

FRUITS AND SEEDS

Fruits develop following fertilisation of female flowers and are the structures in which a tree or shrub produces its seeds. The appearance of these fruits is as varied as are the trees themselves and a number of ornate, and sometimes colourful, structures are easily recognised. In almost all tree and shrub species, dispersal of the seeds away from the parent plant is essential because it reduces competition for nutrients and available light. The means by which any given species achieves dispersal is often a major influence on the appearance of its fruit, and sometimes of the seeds themselves. Some use wind power and have winged seeds to carry them away, while others surround their fruits by luscious flesh to attract birds or mammals, which eat the fruits without digesting the seeds; these are then spread in their droppings. The seeds of a few species are carried and dispersed by water.

The names by which fruits are known are varied and most people will be familiar with terms such as *berry* and *nut*. In botanical parlance, however, there are a number of other ways in which fruits are described. Confusingly, there are many instances where structures that people refer to as nuts or berries, for example, may have an altogether different classification in the strict botanical sense.

On these pages I have tried to illustrate as wide a selection of fruit structure as space permits, but have chosen to keep the nomenclature simple. However, the following definitions of fruit types may help the reader cross-refer to other, more detailed, tree books:

Achene – a dry, indehiscent, one-seeded fruit with a papery case.
Berry – a succulent fruit containing several seeds that lack a hard case.
Capsule – a dry, many-seeded, dehiscent fruit.
Cone – a compact, usually woody structure comprising a central axis bearing radiating, seed-bearing tiers.
Drupe – a succulent or spongy fruit, usually one-seeded, the seed with a hard case.
Nut – a dry, indehiscent, one-seeded fruit with a hard, woody case.

CONES AND CONE-LIKE STRUCTURES

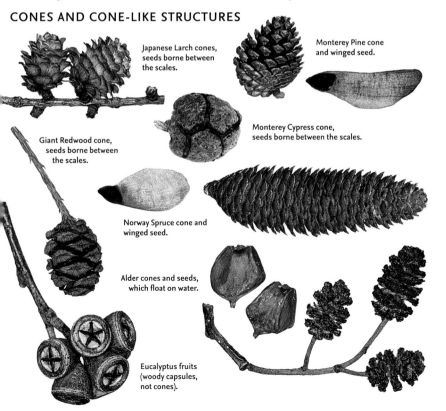

Japanese Larch cones, seeds borne between the scales.

Monterey Pine cone and winged seed.

Monterey Cypress cone, seeds borne between the scales.

Giant Redwood cone, seeds borne between the scales.

Norway Spruce cone and winged seed.

Alder cones and seeds, which float on water.

Eucalyptus fruits (woody capsules, not cones).

WINGED SEEDS

Silver Birch seed (technically, a nut).

Sycamore, paired and winged seeds.

Field Maple, paired and winged seeds.

Ash; a cluster of winged seeds and an individual seed (technically, an achene).

BERRIES AND BERRY-LIKE FRUITS

Hawthorn berries.

Guelder-rose fruits (technically, drupes).

Alder Buckthorn berry.

Whitebeam berries.

Wild Service-tree berries.

Holly fruits (technically, drupes).

NUTS AND NUT-LIKE FRUITS

Sweet Chestnut fruit and individual nut.

Beech fruit, in which nuts are contained.

Hazel nut cluster and individual nut.

Sessile Oak acorn (nut).

Walnut 'nut' (technically, a drupe, the spongy outer layer having been removed here).

Pedunculate Oak acorns (nuts).

Almond 'nut' (technically, a drupe, the spongy outer layer having been removed here).

SUCCULENT AND CLUSTERED FRUITS

London Plane fruits (technically, a cluster of achenes).

Wild Crab.

Cultivated Apple, showing cross section and pips.

HISTORY OF WOODLAND MANAGEMENT

Following the retreat of the glaciers at the end of the last Ice Age, Britain and Ireland soon became vegetated and before long dense forest cloaked the land. However, once man arrived on the scene, this soon began to change. Slowly but surely, wooded areas were cleared to create agricultural land, and to produce timber for fuel and building materials.

Initially, woodland use was probably on a rather *ad hoc* basis. But, in recognition of the value of trees, and the impact that uncontrolled exploitation can produce, most significant areas of woodland have been managed, to varying degrees, at least since medieval times, with the aim of creating a sustainable resource. Although most woodlands are man-made and classified in ecological terms as *semi-natural*, the continuity of woodland cover in relatively stable habitats has had an unintentional, but markedly beneficial, impact on wildlife.

As a general rule, oak (both Pedunculate and Sessile) has always been the timber of choice for building and construction. Consequently, even in managed woods, oaks are often left to reach maturity before being felled, and seedlings are planted to take their place eventually. Typically, in managed oak woodland, the trees are fairly evenly spaced so that, even at a modest size, the canopy is more or less complete but not crowded.

As a very rough guide, the degree to which any given woodland has been managed can be discerned by the proportion of multi-stemmed trees (other than oak) that it contains.

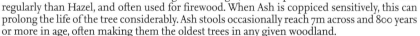

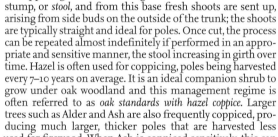

This is the result of man's management over the centuries, with timber harvested periodically without the destruction of the tree or shrub in question; the two familiar techniques are *coppicing* and *pollarding*.

Coppicing involves cutting back the tree or shrub to a stump, or *stool*, and from this base fresh shoots are sent up, arising from side buds on the outside of the trunk; the shoots are typically straight and ideal for poles. Once cut, the process can be repeated almost indefinitely if performed in an appropriate and sensitive manner, the stool increasing in girth over time. Hazel is often used for coppicing, poles being harvested every 7–10 years on average. It is an ideal companion shrub to grow under oak woodland and this management regime is often referred to as *oak standards with hazel coppice*. Larger trees such as Alder and Ash are also frequently coppiced, producing much larger, thicker poles that are harvested less regularly than Hazel, and often used for firewood. When Ash is coppiced sensitively, this can prolong the life of the tree considerably. Ash stools occasionally reach 7m across and 800 years or more in age, often making them the oldest trees in any given woodland.

In southern England in particular, traditional coppicing of Hazel provides ideal conditions for Bluebells to thrive.

Although coppicing is an excellent process for producing regular crops of timber, it has the drawback that, at least initially, the stools are vulnerable to grazing animals and often have to be protected. An alternative is to cut the tree at around head height, so that the sprouting shoots are beyond the reach of deer and cattle; this is called pollarding. Beech, Hornbeam and willows are frequently pollarded but, in some areas, oaks are also subjected successfully to this management regime. Pollarding is often used on isolated trees, as well as in woodland.

Appropriate woodland management, particularly involving coppicing and pollarding, is a relatively time-consuming process and does require a degree of skill on the part of the practitioner. This explains why some landowners neglect their woodlands or manage them insensitively, for example by clear-felling or felling at inappropriate times of year. Fortunately, many coppice and pollard products have value-added appeal and the Forestry Authority provides grants for well-managed woodland where an approved management plan is in place.

Ancient pollarded Hornbeams.

Willows are frequently pollarded to produce slender poles for weaving.

TRADITIONAL TIMBER USES

Visit a rural museum or an antique shop and you will soon appreciate the significance of timber and woodland products to daily life in centuries past. Although the relative importance of trees may have declined in recent years, in rural areas timber products still feature prominently in the local economy. The following are just a few of the ways in which woodland products were, and to some extent still are, used.

BUILDING TIMBER
Timber from both native species of oak has always provided the favoured construction material for building, because of its inherent strength, durability and resistance to decay. Oak is still used today in the construction of traditional timber-frame buildings.

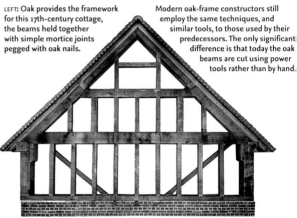

LEFT: Oak provides the framework for this 17th-century cottage, the beams held together with simple mortice joints pegged with oak nails.

Modern oak-frame constructors still employ the same techniques, and similar tools, to those used by their predecessors. The only significant difference is that today the oak beams are cut using power tools rather than by hand.

FUEL
It is likely that timber has been used as a fuel throughout the period of man's occupancy of Britain and Ireland. Given the vagaries of British weather, it is hardly surprising that it was, and still is, used as a source of heating. And, of course, it is also important in cooking, since many of the foods that we traditionally eat are either tastier or more digestible, or both, when cooked as opposed to raw.

ABOVE: Firewood provides the greatest amount of heat if the timber is cut, stacked and seasoned for at least a year before being split for burning. Adequate seasoning ensures that the wood is dry and has lost some of its oils and resins.

RIGHT: Almost all tree species can be burned as firewood although some are more calorific than others. Discerning firewood enthusiasts favour timber from birches, Ash, Alder, Beech and extremely well seasoned oak.

Traditional charcoal burns were performed in covered pits but today purpose-built kilns are used (*left*). However, despite modern technology, creating a good charcoal burn is still more of an art than a science. Charcoal is carbonised wood (*right*), which burns at a much higher temperature than timber itself. In the past it was important in smelting iron and in other manufacturing processes. It is also the main ingredient in gunpowder: charcoal from Alder and Alder Buckthorn was particularly favoured for this purpose.

HOUSEHOLD PRODUCTS

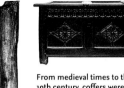

From medieval times to the 19th century, coffers were all the rage for storing clothes and linen. This 17th-century Welsh example (*above*) is made from Sessile Oak.

Oak nail.

Oak (shown) and elm planks were widely used as floorboards (*above*) until the 19th century but were gradually replaced by cheaper softwood alternatives.

This 17th-century oak panelling (*above*) covers up an otherwise bare and uninteresting wall. The frames that surround the individual panels are held together by simple joints and pegged with oak nails.

English coffers (*below*) range in style and complexity from simple plank-constructed blanket boxes (A) to more elaborate panelled and inlaid affairs (B).

Beech timber is easily turned on a pole lathe and was widely used to make chairs (*right*). Unfortunately, however, it is extremely prone to woodworm attack.

Freshly cut willow wands can be woven to make baskets (*right*) and other products. Wands can also be dyed for added decorative effect.

This rustic barrel (*above*) was created by steam-bending a relatively thin strip of Ash. It has stood the test of time, having lasted for more than 100 years.

Large barrels are traditionally made from oak but in this instance a miniature version has been crafted from Scots Pine (*right*).

Turned Beech has been used to make a screw-top pot (*left*), both the top and the pot itself having been turned from the same piece of wood.

Box is a durable wood and easily turned, making it useful as a handle for implements such as this corkscrew (*left*).

The various species and hybrids of poplar are used to make packing crates, and also matches (*left*).

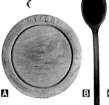

This turned wooden cup (*left*) is made from Pear and is waxed and stained with much use.

Sycamore (*above*) does not stain easily, nor does it taint food. Consequently, it is often used to make kitchen products such as bread boards (A), spoons (B) and rolling pins (C).

For rural people in the past, clogs (*above*) were a much-needed means of ensuring dry feet. Typically they were made from Alder, which has low heat conductivity and good resistance to water.

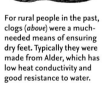

A traditional Sussex trug basket (*right*) is made from thin strips of Sweet Chestnut, creating an item that is both decorative and useful.

Besom brooms (*right*) are made from different materials in various parts of the country, but birch branches are most widely used.

Cricket bats (*right*) are made from a single section of Cricket-bat Willow, originally roughly triangular in cross section, with a cane handle.

HOW TO USE AN OAK TREE

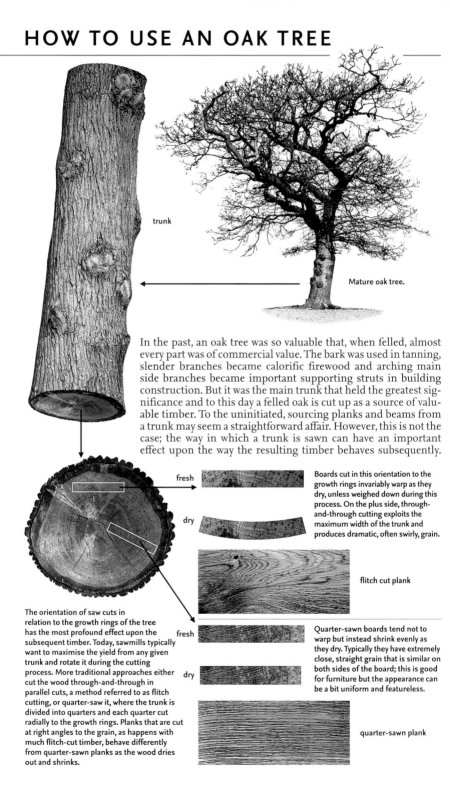

trunk

Mature oak tree.

In the past, an oak tree was so valuable that, when felled, almost every part was of commercial value. The bark was used in tanning, slender branches became calorific firewood and arching main side branches became important supporting struts in building construction. But it was the main trunk that held the greatest significance and to this day a felled oak is cut up as a source of valuable timber. To the uninitiated, sourcing planks and beams from a trunk may seem a straightforward affair. However, this is not the case; the way in which a trunk is sawn can have an important effect upon the way the resulting timber behaves subsequently.

fresh

dry

Boards cut in this orientation to the growth rings invariably warp as they dry, unless weighed down during this process. On the plus side, through-and-through cutting exploits the maximum width of the trunk and produces dramatic, often swirly, grain.

flitch cut plank

The orientation of saw cuts in relation to the growth rings of the tree has the most profound effect upon the subsequent timber. Today, sawmills typically want to maximise the yield from any given trunk and rotate it during the cutting process. More traditional approaches either cut the wood through-and-through in parallel cuts, a method referred to as flitch cutting, or quarter-saw it, where the trunk is divided into quarters and each quarter cut radially to the growth rings. Planks that are cut at right angles to the grain, as happens with much flitch-cut timber, behave differently from quarter-sawn planks as the wood dries out and shrinks.

fresh

dry

Quarter-sawn boards tend not to warp but instead shrink evenly as they dry. Typically they have extremely close, straight grain that is similar on both sides of the board; this is good for furniture but the appearance can be a bit uniform and featureless.

quarter-sawn plank

There are few finer testaments to the strength and durability of oak than the timber-framed Great Barn, in Old Basing, Hampshire. The main timbers came from oaks felled in 1534. The curved side supports would have come straight from the side branches of oaks such as the one on the facing page.

RECOGNISING TIMBER

Wood is a wonderfully versatile, natural product and one that has served the purposes of man since time immemorial, the timbers of different tree species often being put to extremely specific, and often entirely different, uses. Each individual tree will have experienced subtly different growing conditions from its neighbours. Unsurprisingly, therefore, the resulting grain structure differs from one tree to another and it even varies within any given tree: look along the length of an oak floorboard for evidence of this. The result can be rather confusing to the untrained eye, with no single piece of timber resembling another. However, with a bit a practice, factors in common can be discerned in timbers from the same tree species. Discounting imports from abroad, the following timber samples illustrate the range of patterns and colours found in native and naturalised timbers widely used in wood-turning and in the manufacture of traditional furniture in Britain and Ireland.

COMMON LIME
Pale yellow with a straight and extremely fine grain. Easily carved and resistant to warping or splitting when dry.

BEECH
Pinkish buff with a straight grain, even texture and the diagnostic presence of small, evenly spaced dark flecks (rays).

HOLLY
Creamy white, sometimes with a slight greenish sheen and a fine but rather irregular grain. Prone to splitting while drying.

SYCAMORE
Creamy white but often grading to pale grey, with rather straight, wide grain. Used for kitchen implements and violin backs.

PEDUNCULATE OAK
Yellowish brown and straight-grained to swirly depending on the cut. Silvery rays are seen on quarter-sawn planks. Hard and durable when dry, easier to work when 'green'.

WILD CHERRY
Pinkish brown, often with a slight golden sheen. The grain is fine and the timber is prized for wood-turning.

ASH
Creamy buff in colour, with widely spaced grain. Flexible, tough and resistant to splitting and splintering.

NORWAY MAPLE
Yellowish to greyish brown with a straight grain. Can be easily steam-bent. Durable when dry.

BOX
Typically yellow with a rather straight, close grain and often some darker marbling. Liable to split while drying but hard and dense when dry.

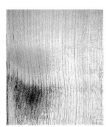

SILVER BIRCH
Creamy white to pale buff with a straight grain. Spalted birch timber (attacked by fungi; *see* p. 312) acquires a beautiful swirly pattern.

LABURNUM
Yellowish brown heartwood contrasts with pale sapwood and this makes it a popular wood-turning timber. Hard and durable when dry.

ASPEN
Pale buff with rather faint, and usually straight, grain. Large pores make it difficult to plane smooth.

BLACKTHORN
Pinkish to yellowish brown with a straight, relatively fine grain. Its colour and texture make it popular for wood-turning.

LONDON PLANE
Ground colour buffish yellow but with distinctive dark flecks that make this an extremely decorative timber.

ENGLISH ELM
Yellowish brown, often with a swirly grain (depending on the cut) that furniture-makers find attractive. The timber is resistant to water and was once used for piping.

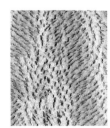

EUROPEAN LARCH
Reddish brown but easily stained darker with wear and tear. Rather straight-grained and knotty; used for flooring and posts etc.

ALDER
Recently cut timber turns bright orange but this fades to pinkish brown with time. Grain rather straight. Timber good for 'green' wood-turning.

DOUGLAS FIR
Reddish brown with a straight to slightly wavy grain, depending on the cut. It is lightweight and used for structural timber and veneers.

PLACES TO VISIT

Britain and Ireland are among the least wooded countries in Europe, especially in terms of natural and completely untouched forest. However, of the ancient semi-natural woodlands that do remain, many are superb and are rich in wildlife, as well as being fine examples of the way in which these habitats can maintain their biodiversity while being exploited commercially on a sustainable basis.

For some tree enthusiasts, visiting their local woodland is enough to satisfy their interest and most people do not have to travel far to achieve this end. However, once in a while it is worth travelling further afield for a change of scenery and species. A visit to one of the splendid arboreta that are scattered around the country will also be a rewarding experience. The following sites (with website addresses) provide some splendid opportunities for woodland walks, visits to superb tree collections, or both.

1. Epping Forest – Extensive semi-natural woodland with ancient pollards and maiden trees. www.cityoflondon.gov.uk/Corporation/living_environment/open_spaces/epping_forest.htm
2. Royal Botanic Gardens, Kew – A huge range of splendid trees, on the edge of central London. www.rbgkew.org.uk
3. Burnham Beeches, Buckinghamshire – One of the best examples of ancient semi-natural woodland in the region, with many veteran pollards. www.cityoflondon.gov.uk/Corporation/living_environment/open_spaces/burnham.htm
4. Windsor Forest, Berkshire – Crown Estate land with ancient trees and working forestry plantations. www.theroyallandscape.co.uk
5. Bedgebury Pinetum, Kent – The finest collection of conifers in the world. www.bedgeburypinetum.org.uk
6. Winkworth Arboretum, Surrey – Peaceful woodland and a fine tree collection. www.nationaltrust.org.uk/main/w-winkwortharboretum.htm
7. Harewood Forest, Hampshire – The largest area of woodland in Hampshire outside the New Forest. www.testvalley.gov.uk/tvlcp/vol1_lca6d.html
8. The New Forest, Hampshire – Ancient hunting forest, now a mosaic of ancient semi-natural woodland, mature plantations and heath. www.hants.org.uk/newforest
9. The Sir Harold Hillier Gardens, Hampshire – The greatest collection of hardy trees in the world. www.hillier.hants.gov.uk
10. Westonbirt Arboretum, Gloucestershire – World-famous tree collection. www.forestry.gov.uk/westonbirt
11. Haldon Forest, Devon – Rolling, wooded hills with both conifer plantations and native woodland. www.haldonforestpark.org.uk
12. Forest of Dean, Gloucestershire – A mixture of mature plantations and ancient semi-natural woodland. www.forestry.gov.uk/website/recreation.nsf/LUWebDocsByKey/EnglandGloucestershireForestofDean
13. National Botanic Garden of Wales – native and introduced trees including an important collection of *Sorbus*. www.gardenofwales.org.uk
14. Gwydyr Forest, Conwy, Wales – Mature conifer plantations and remnant pockets of semi-natural Sessile Oak woodland. www.forestry.gov.uk/website/wildwoods.nsf/SearchAgentView/WalesConwyNoForestGwydyr
15. National Botanic Gardens of Ireland – an excellent collection of native and introduced trees. www.botanicgardens.ie
16. Vale of Clara Nature Reserve, Wicklow, Ireland – Extensive areas of semi-natural oak woodland. www.npws.ie/NatureReserves/Wicklow
17. Wicklow Mountains National Park, Ireland – Contains pockets of ancient semi-natural oak woodland and superb ground flora. www.npws.ie/NationalParks/WicklowMountainsNationalPark
18. Killarney National Park, Ireland – Arguably the finest ancient semi-natural Sessile Oak woodland in Ireland. www.npws.ie/NationalParks/KillarneyNationalPark
19. Breen Wood, Antrim, Northern Ireland – Extensive semi-natural Sessile Oak and Downy Birch woodland. www.jncc.gov.uk/ProtectedSites/SACselection/SAC.asp?EUCode=UK0030097
20. Brodick Castle, Garden and Country Park, Isle of Arran, Scotland – for *Sorbus arranensis* and *S. pseudofennica*. www.nts.org.uk

21. Abernethy and Rothiemurchus Forests, Aviemore, Scotland – Ancient Caledonian Pine forest. www.rspb.org.uk/reserves/guide/a/abernethyforest/index.asp
22. Tay Forest Park, Highlands, Scotland – Extensive mixed and conifer woodland. www.forestry.gov.uk/website/Recreation.nsf/LUWebDocsByKey/ScotlandPerthand KinrossTayForestPark
23. Kielder Forest, Northumberland – A vast area of mature conifer plantation. www.northumberland.gov.uk/VG/kielder.htm
24. Dalby Forest, Yorkshire – Semi-natural woodland and mature conifer plantations in the south of the North York Moors National Park. www.forestry.gov.uk/dalbyforest
25. Sherwood Forest, Nottinghamshire – An ancient woodland that is both a country park and a national nature reserve. www.nottinghamshire.gov.uk/home/leisure/ countryparks/sherwoodforestcp.htm
26. Thetford Forest, Norfolk – Mature conifer plantation and broadleaved woodland. www.forestry.gov.uk/website/Recreation.nsf/LUWebDocsByKey/ EnglandEastAngliaForestsofEastAngliaThetfordForestPark
27. Cambridge University Botanic Garden – An interesting collection of trees, including many scarce native species. It holds the national collection of *Sorbus*. www.botanic.cam.ac.uk

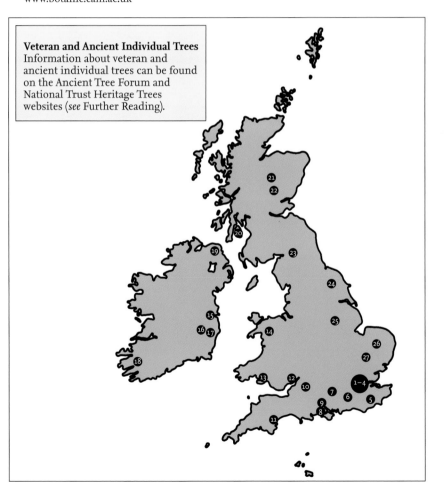

Veteran and Ancient Individual Trees
Information about veteran and ancient individual trees can be found on the Ancient Tree Forum and National Trust Heritage Trees websites (*see* Further Reading).

CONSERVATION

Since man first colonised the British Isles, trees have been viewed as a resource, and treated as such. That is not to say that their significance was not understood, but it is only perhaps in the last two centuries, and particularly in the last 50 years, that they have begun to be appreciated for their intrinsic interest and beauty, and planted to this end.

It is important to remember that almost any British woodland that you come across will, at the very least, have been heavily influenced by man and its current appearance will be a direct result of centuries of management. So it is ironic, perhaps, that many of our finest pockets of semi-natural woodland have also been destroyed by man, grubbed out or clear-felled to gain agricultural land or a quick profit. From an ecological point of view it is not so much the individual trees themselves that are important but the woodland habitat that they collectively form. In Britain and Ireland today, our ancient semi-natural woodland is just a tiny fraction of what once cloaked the land, and especially in the last 50 years the loss and degradation has continued apace. Some of our best woodlands are protected by nature reserve status, for what that's worth, but a significant number remain in private hands and are depressingly unprotected. Of course, enlightened landowners do appreciate their woodland, often for its contribution to the landscape generally as much as for its wildlife value. However, in many cases, woodland is seen as a resource pure and simple. Its continuing economic value is what saves it from destruction, and it is the need to obtain tree-felling licences, rather than any conservation law, that controls its exploitation.

For me, considering how little we have left, the loss or degradation of any more ancient semi-natural woodland is wanton vandalism. To combat future decline, I would urge anybody interested in trees to donate as much money as possible to conservation bodies for the purchase of land. In this context, the only organisations that I would consider donating money to personally are the Wildlife Trusts, the Woodland Trust, the National Trust and the Royal Society for the Protection of Birds.

Even today, while most appreciate the beauty of trees, not everybody respects or understands their role in ecology, or their significance to native wildlife. I hope that *Complete British Trees* will instil in its readers a greater respect for the importance of British and Irish trees.

ABOVE: Herb-Paris is a floral indicator that the woodland in which it grows is ancient.

OPPOSITE PAGE: This massive coppiced Ash stool is probably more than 600 years old.

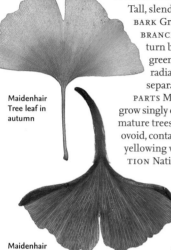

Maidenhair Tree *Ginkgo biloba* (Ginkgoaceae) 28m

Tall, slender to slightly conical deciduous tree with one main trunk. BARK Grey-brown, corky and deeply ridged in mature trees. BRANCHES Spreading with long, greenish-brown shoots, which in turn bear shorter brown shoots. LEAVES Yellowish green to dark green, fan-shaped, divided at least once and up to 10–12cm long; radiating veins reach margins. Leaves on long shoots widely separated; those on short shoots closely packed. REPRODUCTIVE PARTS Male catkins yellow, in small upright clusters; female flowers grow singly on a 5cm-long pedicel. Flowers seldom seen in Britain; most mature trees in the region are male. FRUITS To 3cm long and usually ovoid, containing a single seed inside a harder shell. Green at first, yellowing with age, becoming foul-smelling. STATUS AND DISTRIBUTION Native of Zhejiang Province of China. Endangered in the wild but widely cultivated. COMMENTS Hardy, surviving in city centres, doing best in warmer areas. Does not tolerate pruning and soon dies back if trunk is damaged.

Maidenhair Tree leaf in autumn

Maidenhair Tree leaf

Monkey-puzzle (Chile Pine)
Araucaria araucana (Araucariaceae) 30m

Evergreen, domed to conical tree with a tall cylindrical trunk. BARK Greyish, tough, heavily ridged and wrinkled, with numerous rings of old stem scars. BRANCHES Horizontal or slightly drooping, evenly distributed around trunk. LEAVES 3–5cm long, oval, bright glossy green and scale-like. Tip triangular with a sharp brownish spine. Leaf base overlapping shoot and next leaf; leaves arranged in a dense spiral on shoot. REPRODUCTIVE PARTS Male cones up to 10cm long, in clusters at shoot tips. Female cones rounded, up to 17cm long and green for first 2 years, growing on upper surface of shoots; large scales tapering to a slender, outwardly curved point, and concealing edible brown seeds 4cm long. Trees are either male or female. STATUS AND DISTRIBUTION Native of the mountains of Chile and Argentina, first brought to Europe in 1795. Now common as an ornamental tree in parks and gardens. Grows well in towns, but prefers well-drained soils.

Monkey-puzzle foliage

Monkey-puzzle cone

Norfolk Island Pine *Araucaria heterophylla*
(Araucariaceae) 15m

Palm-like evergreen. BARK Becoming scaly. BRANCHES Growing up full extent of trunk; horizontal but with upswept shoots. LEAVES Scaly, on young plants open and spreading, showing the shoot they are growing on; on older trees closely packed, incurved and hiding the shoot. REPRODUCTIVE PARTS Trees are either male or female, and it is not possible to determine which is which until they flower, something that seldom happens in Britain. STATUS AND DISTRIBUTION Native to Norfolk Island (N of New Zealand). Thrives outdoors only in the extreme south-west of Britain.

Maidenhair Tree bark

Maidenhair Tree foliage

Maidenhair Tree autumn

Monkey-puzzle bark

Monkey-puzzle foliage

Monkey-puzzle

Norfolk Island Pine leaf

Norfolk Island Pine foliage

Common Yew *Taxus baccata* (Taxaceae) 25m

Common Yew leaf and fruit

Broadly conical conifer with dense foliage. The bole of a mature tree may be long and twisted. BARK Reddish, peeling to reveal reddish-brown patches beneath. BRANCHES Level or ascending, with an irregular pattern of slightly pendulous twigs growing from them. LEAVES Flattened, needle-like, up to 4cm long and 3mm wide, narrowing to a sharp point. Dark glossy green above and paler below with 2 pale yellowish bands. Leaves arising spirally around the twig but flattened to lie in a row on either side of the twig. REPRODUCTIVE PARTS Male and female flowers on separate trees. Male flowers solitary, comprising clusters of yellowish anthers that release a fine dust of pollen. Female flowers mostly solitary, greenish and giving rise to hard fruits each surrounded by a bright red fleshy aril with a depression at tip. STATUS AND DISTRIBUTION Native of much of Europe (including Britain), NW Africa and SW Asia, usually preferring drier lime-rich soils; ancient Common Yew woods occur in S England. COMMENTS Very poisonous to humans and livestock. Widely planted for ornament in its many cultivated forms. It is the subject of myths and superstitions and very long-lived: many specimens in old country churchyards are more than 1,000 years old. Quite tolerant of exposure to harsh weather and atmospheric pollution in towns, and can also grow in shady places. Timber (particularly when burred) is used as a decorative veneer and in the past was used to make longbows.

Common Yew fruit

TOPIARY

Common Yew is the species that discerning gardeners use to create substantial, impenetrable and long-lasting hedges. It is the tree that responds best to being clipped into ornamental shapes, the practice known as topiary. Common Yew's evergreen foliage forms dense cover, ideal for creating seamless flat surfaces, smooth curves or sharp edges. Also good for hedging, it can be cut back virtually to the trunk, and will still regenerate a thick cover of foliage within a few years.

Clipped Common Yew hedge and topiary

SIMILAR TREES

Irish Yew *T. baccata* 'Fastigiata' (25m) Differs from Common Yew in having a more columnar, upright form with ascending branches. Leaves, flowers and fruits almost identical to those of Common Yew. Present-day plants of this variant are descendants of one of a pair of trees found in County Fermanagh, Ireland, in the mid-18th century.

Japanese Yew *T. cuspidata* (10m) Similar to Common Yew, but leaves more erect on the twigs, and yellow beneath. Fruits in clusters. Thrives where sheltered by other trees.

Common Yew

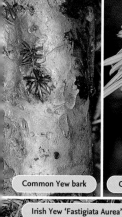

Common Yew bark

Common Yew male flowers

Ancient Common Yew, Kingley Vale

Irish Yew 'Fastigiata Aurea'

Irish Yew foliage

Irish Yew

California Nutmeg needles

California Nutmeg *Torreya californica* (Taxaceae) 20m

Broadly conical tree with a stout bole in mature specimens, producing high-quality timber. BARK Reddish grey with narrow ridges. BRANCHES Long and almost horizontal in mature trees, supporting descending lines of greenish shoots. LEAVES Needle-like with 2 pale greyish bands on the underside; growing in a row on each side of shoot, and smelling of sage if crushed. REPRODUCTIVE PARTS Trees usually either male or female. Male flowers like small yellowish catkins, borne on undersides of shoots. Fruits ovoid, 5cm long, green with purplish streaks; like nutmegs and containing a single (inedible) seed. STATUS AND DISTRIBUTION Native of mountain woodlands in California. Often planted in mature gardens in Britain.

SIMILAR TREE

Japanese Nutmeg *T. nucifera* (12m) Slender and conical with sparse foliage, often looking sickly yellow. Needle-like leaves smelling of sage when crushed. Shoots become reddish brown after 2 years.

Plum Yew needles

Plum Yew (Japanese Cow-tail Pine)

Cephalotaxus harringtonia (Cephalotaxaceae) 6m

Small, bushy, yew-like evergreen tree. LEAVES Leathery, spineless, in dense clusters on slightly downcurved twigs. *C. harringtonia* var. *drupacea* is a more frequently seen variant with shorter leaves growing almost vertically on gracefully curving shoots and showing their silvery-green lower surfaces. *C. harringtonia* 'Fastigiata' is an upright form with much darker foliage; leaves reaching 7cm in length, but shorter near tip of current year's growth. REPRODUCTIVE PARTS Creamy-white male flowers in small clusters on underside of twigs; female flowers (on trees of separate sexes) on very short stalks, later giving rise to small greenish plum-like fruits that ripen reddish. STATUS AND DISTRIBUTION Known only as a garden plant, originally from Japan, and never seen in the wild.

Chinese Plum Yew (Chinese Cow-tail Pine)

Cephalotaxus fortunei (Cephalotaxaceae) 10m

Chinese Plum Yew needles

Small, densely foliaged tree. Usually has a single bole but sometimes 2 or 3. BARK Reddish, peeling. BRANCHES Dense foliage sometimes so heavy that the branches sag. LEAVES Flattened needles up to 10cm long, greenish and glossy above with 2 pale bands on the underside; borne on either side of bright green shoots. REPRODUCTIVE PARTS Yellowish or creamy male and female flowers on separate plants, opening in spring. Fruits up to 2.5cm long, oval, with fleshy, purple-brown skin. STATUS AND DISTRIBUTION Native of mountain forests in central and E China; sometimes grown as an ornamental garden tree. COMMENT Seed was brought to Britain by Robert Fortune (hence *fortunei*) in 1849, and some trees originating from this time are still alive.

Chinese Plum Yew fruit

California Nutmeg bark

California Nutmeg fruits and foliage

California Nutmeg

Plum Yew fruits and foliage

Chinese Plum Yew fruits and foliage

Chilean Plum Yew *Prumnopitys andina* (*Podocarpus andinus*) (Podocarpaceae) 20m

Resembles a yew, but unrelated. Grows either with a single upright bole and horizontal branches, or sometimes with several boles and more upright branches. BARK Dark grey, smooth, with occasional scars and ridges. LEAVES Flattened, needle-like, up to 2.5cm long; deep bluish green on upper surface with 2 pale bands on underside. Leaves much softer than true yew leaves, except in young trees when they are more leathery and bear small spines. Borne in dense shoots, either arranged in 2 ranks on either side of the shoot or spread all round it. REPRODUCTIVE PARTS Male catkins yellow, in branched clusters near ends of shoots. Female flowers greenish, in small spikes at tips of shoots. Flowers on plants of different sexes, opening in spring. FRUITS Like small green plums at first, containing a single seed. May ripen to become blackened and covered with a fine bloom like sloes. STATUS AND DISTRIBUTION Native of the mountains of Argentina and Chile. COMMENTS A fairly hardy tree that is tolerant of clipping and severe pruning to form hedges.

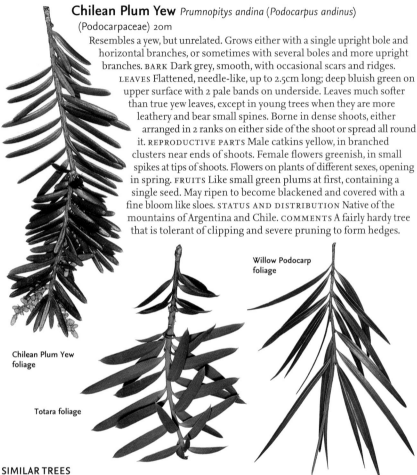

Willow Podocarp foliage

Chilean Plum Yew foliage

Totara foliage

SIMILAR TREES
Totara *Podocarpus totara* (18m) A large tree in its native New Zealand, but in Britain thriving only in sheltered gardens in the west, if protected from frequent frosts. Leaves flattened, tough and leathery, with noticeable spines, appearing yellowish green. Bark greyish brown, peeling off in mature specimens.
Willow Podocarp *Podocarpus salignus* (20m) Sometimes a multi-stemmed bush, but can form a larger tree on a stronger bole, with dark orange-brown bark that peels off in strips from mature specimens. Leaves willow-like, up to 12cm long, with a leathery appearance but a softer, more pliable texture. A native of Chile that yields useful building timber.

Prince Albert's Yew *Saxegothaea conspicua* (Podocarpaceae) 18m

Often resembling a large yew, with a strong ribbed bole. BARK Reddish or purple-brown, peeling off in rounded scales. LEAVES Flattened, curved needles up to 3cm long, arranged untidily on the shoot. Tinged purple at first, becoming greener later. Note 2 pale bands on underside. Crushed leaves smell of grass. REPRODUCTIVE PARTS Male flowers purplish, growing in leaf axils on undersides of shoots. Female flowers small, blue-grey, giving rise to tiny greenish conelets at tips of shoots. STATUS AND DISTRIBUTION Native of forests of S Chile and Argentina, preferring damp, sheltered sites. Best specimens outside native area are found in Ireland and SW England.

Chilean Plum Yew foliage

Chilean Plum Yew

Totara foliage

Totara

Willow Podocarp foliage

Prince Albert's Yew foliage

Willow Podocarp

Leyland Cypress (*Leylandii*) × *Cupressocyparis leylandii* (Cupressaceae) 35m

Evergreen hybrid between the Nootka and Monterey Cypresses (*see* pp. 78–80), first raised in 1888. Normally a tall, narrowly conical tree; densely foliaged, fast-growing and hardy. BARK Reddish brown with thin vertical ridges; usually hidden by dense branches. BRANCHES Arising along whole length of trunk, almost vertical, with a dense growth of green shoots. LEAVES Pointed, scale-like, about 2mm long. REPRODUCTIVE PARTS Male and female cones seldom produced, but occurring on same tree. Male cones small, yellow, at tips of shoots; releasing pollen in March. Female cones up to 3cm across, rounded, with 8 scales each bearing points; green at first, becoming brown and shiny. STATUS AND DISTRIBUTION Man-made hybrid, hence does not have a native range; very widespread in parks and gardens. Tolerant of most soil types. COMMENTS Grows even in town centres and often clipped into hedges, suffering mutilations because of proximity to buildings or excessively vigorous growth. Two cultivars are commonly planted: 'Haggerston Grey', with leaf sprays arising at all angles; and 'Leighton Green', with thicker shoots and leaves in longer, flatter sprays. Golden forms, such as 'Castlewellan', are also widely planted.

Leyland Cypress foliage

Lawson's Cypress *Chamaecyparis lawsoniana* (Cupressaceae) 40m

Erect, narrowly conical evergreen with dense foliage. Sometimes grows as a single slender trunk, but trunk often repeatedly forked. BARK Cracking vertically into long greyish plates.

BRANCHES Trunk bears many small branches, each in turn bearing numerous smaller shoots that are usually flattened and pendulous. LEAVES Small, scale-like, up to 2mm long and flattened along shoot, in opposite pairs, and showing paler colours on underside of shoot. Crushed leaves smell of parsley. REPRODUCTIVE PARTS Male flowers, at tips of twigs, are small reddish cones, up to 4mm long, shedding pollen in March. Female cones, found on same tree, are up to 8mm in diameter when young, greenish blue at first, and becoming yellowish brown with age. Each cone has 4 pairs of scales that have a depression in the centre; they gradually open up to reveal the winged seeds inside the cone. STATUS AND DISTRIBUTION Native of W USA, introduced to Britain in 1854, and now a common park and garden tree in its many cultivars. COMMENTS In its native Oregon and NW California, height can exceed 60m, but British specimens are still growing and, so far, none has greatly exceeded 40m. Today, numerous cultivars exist, with leaf colours ranging from blue-grey through various

Lawson's Cypress foliage

greens to rich gold. Most are narrowly conical with lower branches arising at ground level, and all show flattened scale-like leaves along shoots.

Lawson's Cypress foliage

Lawson's Cypress 'Wisseleii'

Leyland Cypress 'Haggerston Grey'

Lawson's Cypress golden form

Lawson's Cypress

Lawson's Cypress 'Minima Aurea'

Hinoki Cypress foliage

Hinoki Cypress *Chamaecyparis obtusa* (Cupressaceae) 25m

Evergreen; recalls Lawson's Cypress. BARK Reddish and soft.
BRANCHES Mainly level. LEAVES Blunt-pointed, bright green
with white lines below, eucalyptus-scented; in flat sprays.
REPRODUCTIVE PARTS Rounded female cones blue-green at first,
yellowing with age. Male cones small, reddish yellow. STATUS AND
DISTRIBUTION Native of Japan and Taiwan. Introduced to Britain in
1861; grows best in wetter areas. COMMENTS Attains great size and age
in the wild, despite slow growth. Cultivars include golden form,
'Crippsii', and several small forms with dense foliage, e.g. Club-moss
Cypress 'Lycopodioides'. Slow-growing dwarf forms are ideal for small
gardens.

Sawara Cypress *Chamaecyparis pisifera* (Cupressaceae) 24m

Sawara Cypress foliage

Evergreen; recalls Lawson's Cypress but with finer, paler foliage, and
more open crown. BARK Reddish brown, peeling in vertical strips.
BRANCHES Mostly level. LEAVES Scale-like with white marks on
undersides; resinous scent when crushed. REPRODUCTIVE PARTS
Male flowers small brownish cones, female flowers paler brown,
growing in clusters at shoot tips. Wrinkled, pea-like cones 6–8mm
across, hidden among foliage. STATUS AND DISTRIBUTION
Native of mountain woodlands in Japan. COMMENTS Favourite
cultivars include 'Plumosa' and golden-leaved 'Plumosa Aurea'.
Both are broadly conical, growing to tall, columnar trees, the
stout bole dividing about 2m above ground level into 3 or 4
stems. 'Squarrosa' is smaller and more conical with a single,
often drooping, leading shoot, and softer blue-green foliage.

Taiwan Cypress *Chamaecyparis formosensis*
(Cupressaceae) 16m

Distinctive evergreen. BRANCHES Upturned and U-shaped. LEAVES
Greenish bronze without white markings below. REPRODUCTIVE PARTS
Female flowers small, green, partly hidden by leaves. STATUS AND
DISTRIBUTION Native of Taiwan, but rare there, though some large, ancient
(up to 3,000 years) specimens still exist. In Britain, planted since 1910 but
still rare.

Nootka Cypress *Chamaecyparis nootkatensis*
(Cupressaceae) 30m

Evergreen, forming an elegant conical tree.
BRANCHES Slightly upturned with pendulous
shoots. LEAVES Tough, scale-like.
Unpleasant smell when crushed.
REPRODUCTIVE PARTS Male flowers
yellow. Cones blue in first year, ripening
through green to brown. STATUS AND
DISTRIBUTION Discovered near Nootka,
on Vancouver Island; occurs elsewhere in Pacific
north-west. Intolerant of lime-rich soils.
COMMENTS Hardy. Cultivars include 'Pendula',
with upswept branches and pendulous shoots;
and 'Variegata', with golden foliage.

Taiwan Cypress foliage

Nootka Cypress foliage

Hinoki Cypress bark

Hinoki Cypress fruits and foliage

Hinoki Cypress 'Aurea'

Sawara Cypress 'Filifera' foliage

Sawara Cypress 'Plumosa Aurea'

Taiwan Cypress bark

Taiwan Cypress foliage

Nootka Cypress foliage

Monterey Cypress *Cupressus macrocarpa* (Cupressaceae) 36m

Large evergreen, pyramidal when young, domed and spreading when mature. BARK Reddish brown; ridged and scaly with age. BRANCHES Crowded, upright on younger trees, more level and spreading with age. LEAVES Small, scale-like, on stiff, forward-pointing shoots; lemon-scented. REPRODUCTIVE PARTS Male cones yellow, up to 5mm across, produced on tips of shoots behind female cones; females 2–4cm across, rounded and bright green at first, maturing purplish green; each scale has a central point. STATUS AND DISTRIBUTION Native near Monterey, California, where it is now rare and never attains the size it can in W Britain and Ireland. Formerly widely planted in hedgerows and shelter-belts. Golden-foliaged form 'Lutea' is more spreading and tolerant of sea winds.

Monterey
Cypress
foliage

Monterey
Cypress
cone

SIMILAR TREE
Mexican Cypress *C. lindleyi* (25m) Branches spreading, slightly pendulous at tips. Leaves unscented with spreading sharp points. Bark brown, peeling. Female cones up to 1.5cm across, shiny brown when mature. Native of Mexico and Guatemala, introduced first to Portugal and later Goa. Not hardy, mainly coastal SW Britain.

Italian Cypress
foliage

Italian
Cypress
cone

Italian Cypress *Cupressus sempervirens* (Cupressaceae) 22m

Slender, upright evergreen with dense, dark green foliage. Usually columnar, but sometimes broadly pyramidal. BARK Grey-brown and ridged. BRANCHES Strongly upright and crowded, bearing clusters of shoots. Numerous young shoots arising from the leading shoots. LEAVES Dark green, scale-like, no more than 1mm long; unscented. REPRODUCTIVE PARTS Small greenish-yellow male cones up to 8mm across, on tips of side-shoots. Elliptical, yellowish-grey female cones, up to 4cm across and ripening brown, near ends of shoots. STATUS AND DISTRIBUTION Native of mountain slopes in S Europe and Balkans, E to Iran. Most wild trees spreading, but elegant columnar form widely planted elsewhere.

Smooth Arizona Cypress *Cupressus glabra* (Cupressaceae) 22m

British form grows into a neat, ovoid tree with blue-grey foliage, often with white tips. In the wild (in Arizona) it is more spreading. BARK Reddish or purplish; falling away in rounded flakes in older specimens, revealing yellow or reddish patches. LEAVES Greyish green, often with a central white spot; grapefruit-scented. REPRODUCTIVE PARTS Male cones small, yellow, at tips of shoots. Female cones oval, up to 2.5cm across when mature, and greenish brown; central blunt projection on scales. STATUS AND DISTRIBUTION Native to Arizona, planted in Britain for ornament and hedging.

Smooth Arizona Cypress
foliage

Monterey Cypress bark

Monterey Cypress cones and foliage

Monterey Cypress

Italian Cypress bark

Italian Cypress cone and foliage

Italian Cypress

Smooth Arizona Cypress cones and foliage

Western Himalayan (Bhutan) Cypress *Cupressus torulosa*

(Cupressaceae) 27m

Western
Himalayan
Cypress
foliage

Ovoid crown recalls *C. glabra* but with a more open habit. Slender green shoots smell of new-mown grass when crushed. BARK Spirally ridged in older trees. BRANCHES Raised, with descending sprays of looser foliage. LEAVES Tiny, scale-like, unmarked, with minute, incurved points. REPRODUCTIVE PARTS Cones less than 15mm across, each scale with a rounded knob. STATUS AND DISTRIBUTION Native to W Himalayas, grown in a few old gardens in the British Isles. COMMENTS Slow-growing.

SIMILAR TREE

Gowen Cypress *C. goveniana* (24m) Forming a compact column in Britain, but more sparse and spreading in its native California. Numerous short shoots, often growing out of the leading shoot at right angles, covered with small, rounded leaves smelling of lemon and herbs when crushed. Cones shiny red-brown, in small clusters.

Patagonian Cypress *Fitzroya cupressoides* (Cupressaceae) 22m

Patagonian
Cypress
foliage

Densely foliaged evergreen. BARK Reddish brown, peeling away in vertical strips. BRANCHES Thick branches grow from low down on the bole and curve upwards to grow almost vertically, bearing descending masses of shoots. LEAVES Hard, blunt-ended scales, curving outwards away from shoot, with white stripes on both surfaces. REPRODUCTIVE PARTS Sometimes a prolific producer of cones, which are small, rounded and brown, and up to 8mm across. STATUS AND DISTRIBUTION Native of mountains of Chile and Argentina, attaining great size and age. COMMENT Named after Captain Fitzroy of HMS *Beagle*, which conveyed Charles Darwin on his explorations of South America in the 1830s.

Incense Cedar *Calocedrus decurrens* (Cupressaceae) 35m

Incense
Cedar
foliage
and
cones

Elegant columnar evergreen with narrowly rounded crown. BARK Dark, cracked into large reddish-brown flakes. BRANCHES Numerous short upright branches running up the trunk from near ground level. LEAVES Scale-like, in whorls of 4, each bearing a short, incurved, pointed tip, adpressed and concealing shoots; smelling of turpentine when crushed. REPRODUCTIVE PARTS Male cones up to 6mm across, ovoid, deep yellow, at tips of lateral shoots. Female cones 2–3cm across when mature, oblong to ovoid and pointed, with 6 scales; 2 large fertile scales have outwardly pointed tips. STATUS AND DISTRIBUTION Native of California and Oregon. Very popular ornamental tree in the British Isles. COMMENTS Easy to propagate from cuttings or from seed.

SIMILAR TREE

Chilean Incense Cedar *Austrocedrus chilensis* (15m) Less regular in outline than Incense Cedar. Sprays of foliage very flattened, not always showing the white stripes seen in Incense Cedar. More tender and shorter-lived than Incense Cedar, thriving only in the west of Britain and Ireland.

Chilean Incense
Cedar foliage

Western Himalayan Cypress foliage

Patagonian Cypress bark

Patagonian Cypress foliage

Western Himalayan Cypress

Incense Cedar

Incense Cedar bark

Incense Cedar cones and foliage

Chilean Incense Cedar bark

Chilean Incense Cedar foliage

JUNIPERS *JUNIPERUS* (FAMILY CUPRESSACEAE)

A widespread group across most of the northern hemisphere from the Arctic to desert regions nearer the equator. Seeds are borne in cones in which the scales become fleshy and merge to look berry-like. All junipers are slow-growing, and none reaches a great size. Many are aromatic.

Common Juniper foliage

Common Juniper *Juniperus communis* (Cupressaceae) 6m

Often no more than a small aromatic evergreen shrub, but occasionally a small bushy tree, the size and shape influenced in part by location and climate. BARK Reddish brown, peeling off in thin sheets on older trees. BRANCHES Mainly upright, bearing 3-angled, ridged twigs. LEAVES Small, pointed, needle-like, in whorls of 3, emerging at right angles from twigs. Leaves up to 2cm long and 2mm wide, with a pale white band on the upper surface. Stalkless, with a thin keel on the lower surface. Crushed foliage has a smell like gin to some people, or apples to others. REPRODUCTIVE PARTS Male cones small, yellow, globular, borne singly in the leaf axils, and producing pollen in March. Female cones rounded, up to 9mm long, green at first, ripening through blue-green to black in the second year when fully ripened. Cones usually contain 3 seeds. Male and female flowers normally on separate trees. STATUS AND DISTRIBUTION In the wild, possibly the most widespread tree species in the world, occurring all round the northern hemisphere from the Arctic tree-line to the drier regions of the Mediterranean, and growing high on mountains, beneath pines, and on limestone grassland. Sometimes exists only in a prostrate form or as an undershrub, especially in extreme conditions, but can grow to a larger tree in the open. Common Juniper is a plant of chalk downland in S England, where it is only very locally common, and often forms a sizeable shrub; further N and W, it is found on open moorland and limestone crags, often growing in its prostrate form. It also occurs as an undershrub in mature native Caledonian pinewoods in Scotland. It is popular in gardens in a variety of cultivated forms. COMMENTS The ripe cones are collected to provide the flavouring for gin and in cooking, and are also eaten by birds. The aromatic wood is used for smoking food and for wood-carving.

Common Juniper dried 'berries' (cones) for culinary use

Common Juniper prostrate form

Common Juniper bark

Common Juniper

Common Juniper cones and foliage

Common Juniper, prostrate form, cones and foliage

Common Juniper foliage

Prickly Juniper foliage

Prickly Juniper (Cade) *Juniperus oxycedrus* (Cupressaceae) 14m

Spreading evergreen shrub or small untidy tree. BARK Brown, sometimes tinged with purple, peeling away in vertical strips. LEAVES Sharply pointed needles arranged in whorls of 3; upper leaf surface with 2 pale bands separated by slightly raised midrib, and lower surface with pronounced midrib. REPRODUCTIVE PARTS Female cones rounded or pear-shaped, maturing to a reddish colour. STATUS AND DISTRIBUTION Native of S Europe, generally preferring dry habitats. Very variable, with 3 subspecies recorded: ssp. *oxycedrus* has leaves 2mm wide and cones no more than 1cm across (favours dry mountains and stony areas); ssp. *macrocarpa* has leaves 2.5mm wide and cones 1.5cm across (favours rocky areas and maritime habitats across S Europe); ssp. *transtagana* has narrower leaves no more than 1.5mm wide and cones less than 1cm across (restricted to maritime sands of S Portugal). COMMENTS A medicinal oil, used to treat skin problems, can be extracted from the wood. Berries can also be used to flavour gin.

Phoenician Juniper foliage

Phoenician Juniper *Juniperus phoenicia* (Cupressaceae) 8m

Small evergreen tree or spreading shrub. LEAVES Scaly twigs bear 2 types of leaves. Young leaves up to 1.5cm long and 1mm wide, sharply pointed, showing pale bands on both surfaces; in bunches of 3 spreading at right angles. Mature leaves only 1mm long, resembling tiny green scales clasping the twig. REPRODUCTIVE PARTS Male cones inconspicuous, at ends of shoots; female cones up to 1.4cm long, rounded and ripening from black through yellowish green to deep red in second year. STATUS AND DISTRIBUTION Native of Mediterranean coasts and Atlantic shores of Portugal. Found in old collections in Britain.

Syrian Juniper *Juniperus drupacea* (Cupressaceae) 18m

Shapely evergreen, forming a slender, tall column of compact, bright green foliage. Occasionally the trunk and crown divide to make a more conical tree. BARK Orange-brown, peeling away in thin shreds. LEAVES Needle-like, pointed with a spine and 2 pale bands on the underside; longer than any other juniper at 2.5cm. Needles in bunches of 3. REPRODUCTIVE PARTS Male trees produce tiny, bright yellowish-green, oval flowers. Female trees produce tiny green flowers in small clusters at tips of twigs, opening in spring, and these develop into rounded, woody cones, about 2cm in diameter, which turn purple-brown when mature. STATUS AND DISTRIBUTION Native of mountain forests in W Asia; range just extends into Greece. Occasionally planted in British and Irish gardens, where it makes a fine specimen tree, tolerating a wide range of soils, but rarely producing cones.

Syrian Juniper foliage

Prickly Juniper foliage

Prickly Juniper cones and foliage

Phoenician Juniper cones and foliage

Syrian Juniper

Syrian Juniper bark

Syrian Juniper foliage

Syrian Juniper foliage

Pencil
Cedar
foliage

Pencil Cedar (Eastern Red Cedar) *Juniperus virginiana*
(Cupressaceae) 17m

Slender pyramidal or sometimes narrowly columnar evergreen, usually with a single trunk. BARK Reddish brown, peeling in vertical strips. BRANCHES Numerous, small and ascending, bearing fine, rounded, scaly twigs. LEAVES Young leaves (in pairs at ends of shoots) needle-like, finely pointed and up to 6mm long; upper surface with bluish band and lower surface green. Mature leaves 1.5mm long, rounded, scale-like, usually growing close to shoot; in various shades of green. Crushed foliage smells of paint. REPRODUCTIVE PARTS Male cones small, yellow, at tips of shoots. Female cones oval, up to 6mm long, ripening in first year, maturing through bluish green to violet-brown. STATUS AND DISTRIBUTION Native of E USA in a variety of habitats. Infrequently planted in Britain and Ireland.

Chinese Juniper *Juniperus chinensis* (Cupressaceae) 18m

Large evergreen with dark green foliage and a sparse habit when mature. BARK Reddish brown, peeling in vertical strips. BRANCHES Level to ascending. LEAVES Young leaves needle-like, 8mm long with sharply pointed tips and 2 bluish stripes on upper surface; mostly in clusters of 3 at base of adult shoots, radiating at right angles. Adult leaves small, scale-like, closely adpressed to shoot. Crushed leaves smell of cats. REPRODUCTIVE PARTS Male cones small, yellow, on tips of shoots. Female cones rounded, up to 7mm long, bluish white at first, ripening purplish brown in second year. STATUS AND DISTRIBUTION Native of Japan and China, often planted in Britain in parks, gardens and churchyards. COMMENTS A popular golden-leaved cultivar, 'Aurea', forms a neat column of golden-green foliage.

Chinese Juniper
foliage

Meyer's Juniper
foliage

SIMILAR TREE
Meyer's Juniper *J. squamata* 'Meyeri' (11m) Small conical evergreen with striking blue-grey foliage when young. Leaves needle-like with paler stripe on underside. Bark of mature trees peeling in thin pinkish-brown scales. Tolerant of poor soils and tough growing conditions, so suited to town gardens.

Drooping Juniper *Juniperus recurva* (Cupressaceae) 14m

Small evergreen with ascending branches but drooping foliage. Outline broadly conical. BARK Greyish brown, peeling in long, untidy shreds. LEAVES Tough, needle-like, clasping shoots; paint-like smell when crushed. REPRODUCTIVE PARTS Male cones yellow, growing in small clusters at tips of shoots. Female cones produced at ends of shoots, becoming oval, black and berry-like when mature, and growing to 8mm across. STATUS AND DISTRIBUTION Native of SW China and the Himalayas; planted in Britain for ornament. COMMENTS Several cultivars exist showing varyingly drooping foliage.

Drooping Juniper
foliage

Chinese Juniper bark

Meyer's Juniper bark and foliage

Pencil Cedar

Chinese Juniper

Pencil Cedar foliage

Chinese Juniper foliage

Drooping Juniper foliage

Meyer's Juniper foliage

Western Red Cedar foliage

Western Red Cedar *Thuja plicata* (Cupressaceae) 45m

Tall, conical tree with buttressed trunk and upright leading shoot. BARK Reddish brown with fibrous plates. LEAVES Tiny, scale-like, clasping shoots in alternate, opposite pairs; glossy, dark green above, paler below with pale markings. Crushed leaves pineapple-scented. REPRODUCTIVE PARTS Male and female cones on separate trees. Small yellow or brownish male cones grow at shoot tips. Female cones ovoid, up to 1.2cm long, with 8–10 spine-tipped scales. STATUS AND DISTRIBUTION Native of W USA, grown in the British Isles for timber or ornament.

Northern White Cedar foliage

Northern White Cedar *Thuja occidentalis* (Cupressaceae) 20m

Broadly conical tree. BARK Orange-brown, peeling in vertical strips. LEAVES Flattened, fern-like sprays of foliage showing white, waxy bands below. Crushed leaves smell of apple and cloves. REPRODUCTIVE PARTS Male cones similar to Western Red Cedar; female cones with rounded tips to cone scales. STATUS AND DISTRIBUTION Native of E North America. Does not thrive in Britain. COMMENTS Many cultivars exist.

Oriental Thuja foliage

Japanese Thuja *Thuja standishii* (Cupressaceae) 22m

Broadly conical tree. BARK Reddish brown, peeling in strips or broader flakes. BRANCHES U-shaped with pendent grey-green shoot tips. LEAVES Tiny, scale-like, on flattened sprays, lemon-scented when crushed. REPRODUCTIVE PARTS Male flowers at shoot tips, dark red at first, yellower when open. Female flowers greenish, in separate clusters on tips of different shoots on same tree; ripening to red-brown, scaly cones. STATUS AND DISTRIBUTION Native of Japan, planted in Britain for ornament.

Japanese Thuja foliage

Oriental Thuja *Platycladus orientalis* (Cupressaceae) 16m

Foliage in flat vertical sprays; both surfaces same shade of green. LEAVES Tiny, scale-like, unscented. REPRODUCTIVE PARTS Male flowers small, yellow-orange; on ends of shoots. Female flowers greenish, becoming cones with prominent hooked scales. STATUS AND DISTRIBUTION Native of China, grown in parks and gardens in the British Isles.

SIMILAR TREE
Korean Thuja *Thuja koraiensis* (11m) The only evergreen with flat fern-like foliage sprays that are completely silvery or white below.

Hiba *Thujopsis dolabrata* (Cupressaceae) 20m

Single-boled conical tree or broad shrub on a divided trunk. LEAVES Scale-like, glossy green above with white bands below and a pointed, curved tip. Leaves clasping shoots, in opposite pairs on flat sprays. REPRODUCTIVE PARTS Small blackish male cones at shoot tips. Rounded female cones singly on ends of shoots on same tree. Mature cones about 1.2cm long, brown. STATUS AND DISTRIBUTION Native of Japan, planted here for ornament. Prefers wet regions with damp soils.

Hiba foliage

Western Red Cedar bark

Northern White Cedar foliage

Japanese Thuja bark

Japanese Thuja foliage

Western Red Cedar

Northern White Cedar

Oriental Thuja cones and foliage

Hiba foliage

Wellingtonia (Giant Sequoia) *Sequoiadendron giganteum*
(Taxodiaceae) 50m

An outstandingly large evergreen in its native California, and now the tallest tree even in many areas of Britain. Forms a striking, narrowly conical tree with a huge tapering bole, ridged and fluted at the base. BARK Thick, spongy and rich red. BRANCHES May not start for several metres above the ground. Lower branches pendulous, upper branches more level. LEAVES Scale-like, green, up to 1cm long, clasping shoots. Aniseed-scented when crushed. REPRODUCTIVE PARTS Small yellow male cones, sometimes abundant, at tips of shoots, releasing pollen in spring. Female cones solitary, sometimes paired; ovoid, up to 8cm long and 5cm in diameter when ripe, deep brown with a corky texture. STATUS AND DISTRIBUTION A native of the Sierra Nevada in California, where it grows in groves on the western slopes of the mountains. First discovered in 1852, it was soon introduced into Britain, where it thrives best in the west. It grows only slowly in polluted air. COMMENTS The thick, spongy bark often develops deep holes that are used by roosting birds, particularly Treecreepers.

Wellingtonia bark

Wellingtonia cone

Wellingtonia foliage

Coastal Redwood foliage

Coastal Redwood *Sequoia sempervirens*
(Taxodiaceae) 50m

An impressively large evergreen, growing to be the tallest tree in the world in its native California and Oregon. Forms a conical to columnar tree with a tapering trunk arising from a broader, buttressed base. BARK Thick, reddish brown, becoming spongy, eventually deeply fissured and peeling. BRANCHES Mostly arising horizontally or slightly pendulous. LEAVES Green twigs support a unique combination of 2 types of leaves arranged in spirals: leading shoots have scale-like leaves, up to 8mm long, clasping the stem; and side-shoots have longer, flattened, needle-like leaves up to 2cm long, lying in 2 rows. Crushed foliage smells of grapefruit. REPRODUCTIVE PARTS Male and female cones on same tree. Male cones small, yellow, on tips of main shoots, releasing pollen in early spring. Female cones pale brown, growing singly on tips of shoots, becoming 2cm long and ovoid. STATUS AND DISTRIBUTION A native of California and Oregon, growing best in the hills where the permanent sea-mists keep the trees supplied with moisture. Unsurprisingly, the biggest British and Irish specimens are in the west and north of the region.

Coastal Redwood bark

Wellingtonia male flowers

Wellingtonia female flowers

Wellingtonia trunk

Wellingtonia

Wellingtonia foliage

Coastal Redwood

Coastal Redwood foliage

Coastal Redwood cones and foliage

Coastal Redwood foliage

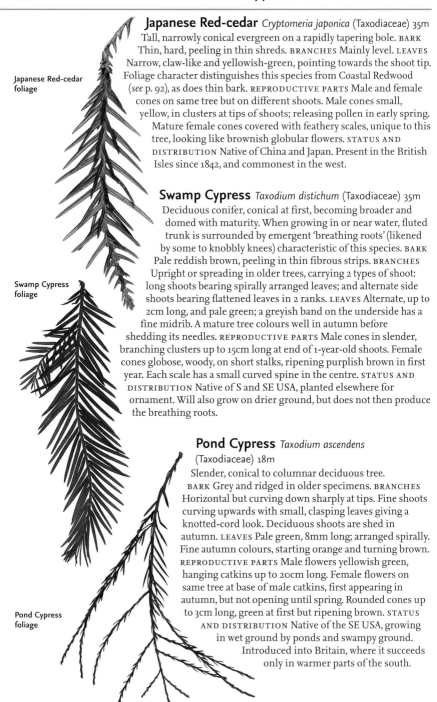

Japanese Red-cedar *Cryptomeria japonica* (Taxodiaceae) 35m

Tall, narrowly conical evergreen on a rapidly tapering bole. BARK Thin, hard, peeling in thin shreds. BRANCHES Mainly level. LEAVES Narrow, claw-like and yellowish-green, pointing towards the shoot tip. Foliage character distinguishes this species from Coastal Redwood (*see* p. 92), as does thin bark. REPRODUCTIVE PARTS Male and female cones on same tree but on different shoots. Male cones small, yellow, in clusters at tips of shoots; releasing pollen in early spring. Mature female cones covered with feathery scales, unique to this tree, looking like brownish globular flowers. STATUS AND DISTRIBUTION Native of China and Japan. Present in the British Isles since 1842, and commonest in the west.

Japanese Red-cedar foliage

Swamp Cypress *Taxodium distichum* (Taxodiaceae) 35m

Deciduous conifer, conical at first, becoming broader and domed with maturity. When growing in or near water, fluted trunk is surrounded by emergent 'breathing roots' (likened by some to knobbly knees) characteristic of this species. BARK Pale reddish brown, peeling in thin fibrous strips. BRANCHES Upright or spreading in older trees, carrying 2 types of shoot: long shoots bearing spirally arranged leaves; and alternate side shoots bearing flattened leaves in 2 ranks. LEAVES Alternate, up to 2cm long, and pale green; a greyish band on the underside has a fine midrib. A mature tree colours well in autumn before shedding its needles. REPRODUCTIVE PARTS Male cones in slender, branching clusters up to 15cm long at end of 1-year-old shoots. Female cones globose, woody, on short stalks, ripening purplish brown in first year. Each scale has a small curved spine in the centre. STATUS AND DISTRIBUTION Native of S and SE USA, planted elsewhere for ornament. Will also grow on drier ground, but does not then produce the breathing roots.

Swamp Cypress foliage

Pond Cypress *Taxodium ascendens*
(Taxodiaceae) 18m

Slender, conical to columnar deciduous tree. BARK Grey and ridged in older specimens. BRANCHES Horizontal but curving down sharply at tips. Fine shoots curving upwards with small, clasping leaves giving a knotted-cord look. Deciduous shoots are shed in autumn. LEAVES Pale green, 8mm long; arranged spirally. Fine autumn colours, starting orange and turning brown. REPRODUCTIVE PARTS Male flowers yellowish green, hanging catkins up to 20cm long. Female flowers on same tree at base of male catkins, first appearing in autumn, but not opening until spring. Rounded cones up to 3cm long, green at first but ripening brown. STATUS AND DISTRIBUTION Native of the SE USA, growing in wet ground by ponds and swampy ground. Introduced into Britain, where it succeeds only in warmer parts of the south.

Pond Cypress foliage

Japanese Red-cedar cones and foliage

Japanese Red-cedar flowers and foliage

Swamp Cypress foliage

Swamp Cypress autumn colours

Pond Cypress bark

Pond Cypress foliage

Pond Cypress

Dawn Redwood *Metasequoia glyptostroboides* (Taxodiaceae) 35m

Conical deciduous conifer with shoots and leaves in opposite pairs. Trunk tapers and is buttressed at base, becoming ridged in older trees. BARK Rich reddish brown, peeling in vertical strips. LEAVES 2.5cm long, flat, needle-like, pale green at first, becoming darker green later; on short lateral shoots that are shed in autumn. Leaves emerge early in spring, turning yellow, pink or red before falling. REPRODUCTIVE PARTS Male and female flowers (seldom seen in Britain) on young shoots in separate clusters on same tree in spring. Males yellow; females greenish, producing rounded cones, green then brown, about 2.5cm across. STATUS AND DISTRIBUTION Native of SW China, unknown as a living tree (known only from fossil records) until 1941. Now a popular garden tree.

LEFT: Dawn Redwood foliage
RIGHT: Dawn Redwood cone

Summit Cedar *Athrotaxis laxifolia* (Taxodiaceae) 10m

Broadly conical evergreen with upcurved branches. BARK Red-brown, fibrous. LEAVES Small, scale-like, up to 6mm long, with sharp tips. Young leaves and shoots bright yellow-green, older shoots and leaves dark green. REPRODUCTIVE PARTS Male and female flowers yellowish brown, in different clusters, on same plant. Cones rounded, 2cm across, green ripening to red-brown, with spiny scales. STATUS AND DISTRIBUTION Native of Tasmania; sometimes planted in Britain.

SIMILAR TREE
King William Pine *A. selaginoides* (20m) Has longer, more curved leaves than Summit Cedar. Foliage resembles a clubmoss. Native of Tasmania, rarely planted here.

King William Pine foliage

Japanese Umbrella Pine *Sciadopitys verticillata* (Taxodiaceae) 23m

Broadly conical evergreen, often with a finely tapering crown, but may be bushy. BARK Red-brown, peeling in long vertical strips. LEAVES Needle-like, up to 12cm long, in umbrella-like clusters. Needles deeply grooved on both sides, dark green above but more yellow below. REPRODUCTIVE PARTS Male flowers yellow, in clusters; female flowers green, at tips of shoots, ripening into ovoid, 7.5cm-long, red-brown cones after 2 years. STATUS AND DISTRIBUTION Native of Japan, but grows well in many parts of Britain and Europe.

Japanese Umbrella Pine needles

Chinese Fir *Cunninghamia lanceolata* (Taxodiaceae) 25m

Broadly conical evergreen conifer with foliage recalling Monkey-puzzle (*see* p. 68). BARK Reddish brown, ridged with age. LEAVES Narrow strap-shaped, pointed, up to 6cm long; glossy green with 2 white bands below. Dead foliage persists inside crown; looks bright orange in sunlight. REPRODUCTIVE PARTS Male and female flowers yellowish; in clusters at shoot tips. Cones rounded, scaly, 3–4cm across, green ripening brown. STATUS AND DISTRIBUTION Native of China; planted in large gardens, mainly in S and W Britain.

Chinese Fir foliage

Dawn Redwood foliage

Japanese Umbrella Pine cone and foliage

Dawn Redwood trunk

Japanese Umbrella Pine bark

Japanese Umbrella Pine foliage

Dawn Redwood

Chinese Fir cones and foliage

Chinese Fir bark

Chinese Fir male cones and foliage

FIRS *ABIES* (FAMILY PINACEAE)

About 50 species, often tall and imposing trees with a pleasing, symmetrical, conical shape on a strong single bole. Cones are erect, and leathery leaves are attached to the shoot by sucker-like structures at the base.

European Silver Fir needles

European Silver Fir *Abies alba* (Pinaceae) 47m

Fast-growing fir, reaching a great size; until 1960s held record for tallest tree in Britain. BARK White on trunk and branches of mature trees, grey on younger trees. LEAVES Thick needles, up to 3cm long, notched at tip and in 2 rows on twigs, which are covered with pale brown hairs. REPRODUCTIVE PARTS Erect cones green at first, maturing orange-brown and up to 20cm long; eventually disintegrating into fan-like scales and toothed bracts, leaving just the protruding woody axis. Cones normally high up on tree. STATUS AND DISTRIBUTION Native of European mountains. At one time widely planted in Britain for timber. Susceptible to aphid attack and vulnerable to late frosts so now largely replaced by species like Noble Fir; large trees now mainly confined to Scotland and Ireland.

Caucasian Fir needles

Caucasian Fir *Abies nordmanniana* (Pinaceae) 42m

Large, shapely fir with thick foliage. BARK Dull grey and fissured with age, forming small square plates. LEAVES Tough, green, forward-pointing needles, in dense rows around brownish twigs; 1.5–3.5cm long, slightly notched at tip and grooved above. REPRODUCTIVE PARTS Male flowers reddish, on underside of shoots. Female flowers greener, upright, in separate clusters on same tree. Cones high up on mature trees (30m); 12–18cm long, dark brown, resinous with projecting, downcurved scales. They break up on the tree. STATUS AND DISTRIBUTION Native from Turkey eastwards. Planted in Britain for ornament.

Delavay's Silver Fir *Abies delavayi* (Pinaceae) 25m

Similar to European Silver Fir, but buds resinous and young twigs smoother, or downy, reddish brown. Cones dark purplish green, shorter, at 10cm, and more rounded, and scales have a long, projecting, sometimes bent, spine. A native of China, and a popular tree in large parks and gardens.

Santa Lucia Fir *Abies bracteata* (Pinaceae) 38m

Tall, narrowly conical evergreen with tapering crown, strong foliage and pointed buds. BARK Black, marked with scars of fallen branches. LEAVES Sharp-spined needles, up to 5cm long, dark green above with 2 light bands below. REPRODUCTIVE PARTS Cones bright green, up to 10cm long, distinctive with long, projecting, hairlike bracts persisting throughout summer. Flowers small and insignificant; males yellowish, growing on underside of shoots, females green, growing on top of shoots. STATUS AND DISTRIBUTION Rare native of S California. Does well in rainier parts of the British Isles.

Santa Lucia Fir bark

Santa Lucia Fir needles

European Silver Fir foliage and flowers

European Silver Fir bark

European Silver Fir foliage

Caucasian Fir foliage

Delavay's Silver Fir cones and foliage

Santa Lucia Fir foliage

European Silver Fir

Santa Lucia Fir

Noble Fir
needles

Noble Fir *Abies procera* (Pinaceae) 50m

Extremely large, narrowly conical conifer when mature. BARK Silver-grey or purplish; developing shallow fissures with age. BRANCHES Youngest twigs reddish brown, hairy, with resinous buds at tip. LEAVES Bluntly pointed needles, 2–3cm long, grooved on upper surface; blue-grey colour is marked by paler bands on both surfaces. REPRODUCTIVE PARTS Male flowers reddish, supported below shoot. Female flowers cylindrical, red or green, resembling small cones, on upper side of shoot; green spine emerging beneath each scale. Cones up to 25cm long, held erect on upper side of branches. They disintegrate in winter, but may be so abundant that branches are damaged by their weight. STATUS AND DISTRIBUTION Native to Pacific NW USA. Planted in Britain since 1850, reaching greatest size in Scotland. COMMENTS Fallen trees provide nurseries for seedlings, which are often found growing in dense lines along lengths of decayed trunk.

Korean Fir
needles

Korean Fir *Abies koreana* (Pinaceae) 15m

Usually broadly conical in outline but sometimes dumpy. BRANCHES Level in conical trees. LEAVES Strap-like, blunt needles, notched at tip and up to 18mm long; dark green above but whitish either side of midrib below. REPRODUCTIVE PARTS Male flowers yellowish; female flowers reddish, maturing into bluish-purple cones that ripen brown. STATUS AND DISTRIBUTION Native of Korea, now widely planted in gardens.

Grecian Fir *Abies cephalonica* (Pinaceae) 36m

Spreading tree. BARK Grey with a hint of orange in young trees; deeper grey and fissured to form squarish plates in maturity. LEAVES Rigid, prickly needles arising from all round hairless red-brown twigs (not in rows); up to 3cm long with 2 white bands below. REPRODUCTIVE PARTS Cones upright, rich golden brown, up to 16cm long. Downcurved triangular bracts protruding from between scales. Mature trees often heavily loaded with cones. STATUS AND DISTRIBUTION Native to Greek mountains. Grows well in dry areas of Britain, but also thrives in wet regions, where it reaches its greatest size.

Grecian Fir
needles

Giant Fir
needles

Giant Fir *Abies grandis* (Pinaceae) 55m

Magnificent when mature. Fast-growing conifer, reaching a height of 40m in as many years. LEAVES Note the comb-like arrangement of soft, shining, green needles in 2 rows on either side of downy olive-green twigs. Needles up to 5cm long, with notched tip and 2 pale bands below; orange-scented when crushed. REPRODUCTIVE PARTS Cones smooth, less than 10cm long and high up on trees at least 50 years old; breaking up on tree to release seeds. STATUS AND DISTRIBUTION Native of coastal W USA. Planted in Britain for ornament and sometimes commercially.

Noble Fir foliage

Noble Fir

Korean Fir cones and foliage

Grecian Fir foliage

Grecian Fir bark

Giant Fir bark

Giant Fir

Giant Fir cones and foliage

Colorado White
Fir needles

Colorado White Fir *Abies concolor* (Pinaceae) 55m

Columnar to conical tree. BARK Dark grey, fissured with age.
BRANCHES Yellowish twigs bear resinous buds. LEAVES Bluish-
grey needles in 2 ranks, curving upwards; to 6cm long with 2 pale
blue bands below. REPRODUCTIVE PARTS Cones cylindrical,
erect, 10cm long, green, ripening purple then brown. STATUS AND
DISTRIBUTION Native of NW USA. Planted elsewhere for ornament.

Alpine Fir
needles

Alpine Fir *Abies lasiocarpa* (Pinaceae) 16m

Narrowly conical tree. BARK Greyish white, smooth with resinous
blisters. LEAVES Notched needles, to 4cm long, greyish green above, 2
white bands below; dense, on upper side of shoot, central ones pointing
forwards. REPRODUCTIVE PARTS Small male flowers yellow, tinged
red; growing below shoot. Female flowers purple, upright; in clusters
on same plant. Cones cylindrical, to 10cm long, purple, ripening brown.
STATUS AND DISTRIBUTION Native of uplands in W USA. COMMENTS
Var. *arizonica* has bluer leaves and corky bark.

Spanish Fir
needles

Spanish Fir (Hedgehog Fir) *Abies pinsapo* (Pinaceae) 25m

Shapely at first, becoming open-crowned and straggly with age. BARK Dark
grey. LEAVES Bluish-grey, usually blunt needles, to 1.5cm long, densely
arranged all around twig. REPRODUCTIVE PARTS Small male flowers red,
opening yellow; female flowers green, in upright clusters above shoot. Cones
cylindrical, tapering, upright and smooth. STATUS AND DISTRIBUTION
Rare native of Sierra Nevada in S Spain. Sometimes planted for ornament
in Britain; tolerates calcareous soils.

SIMILAR TREE

A. × insignis (20m) A hybrid between Spanish and Caucasian Firs, planted
for ornament. Needles blunt, to 3.3cm long; arranged around shoot,
sometimes with a parting below. Twigs hairy. Female cones to 20cm long
and 5cm in diameter, with bracts pointing away from lower few scales.

Pacific Silver
Fir needles

Pacific Silver Fir (Beautiful Fir) *Abies amabilis* (Pinaceae) 32m

Has luxuriant foliage, a strong trunk and thick tapering crown. BARK Silvery.
LEAVES Glossy, to 3cm long, silvery below; densely packed; orange-scented
when crushed. REPRODUCTIVE PARTS Smooth oval cones tinged purple; on
upper surface of twigs. STATUS AND DISTRIBUTION Native of NW USA.
Widely planted for ornament in wet areas of the British Isles.

Veitch's Silver Fir *Abies veitchii* (Pinaceae) 28m

Has fluted and ribbed trunk. BARK Silvery grey and smooth,
becoming scaly with age. LEAVES 3cm long, notched and silvery
below. REPRODUCTIVE PARTS Small male flowers red, looking
yellow as they open. Female flowers greenish red; in upright
clusters. Cones cylindrical, smooth and erect, to 7.5cm long,
purplish, ripening brown. STATUS AND DISTRIBUTION
Native of Japan. Introduced to Britain in 1879; thrives only
in N Scotland.

Veitch's Silver
Fir needles

Colorado White Fir foliage

Alpine Fir foliage

Spanish Fir foliage

Colorado White Fir bark

Colorado White Fir

Pacific Silver Fir cones and foliage

Spanish Fir

Veitch's Silver Fir foliage

CEDARS *CEDRUS* (FAMILY PINACEAE)
A small genus with just 4 species, confined to Europe and Asia. Cedars flower in autumn, produce large woody cones and have strong, aromatic timber. A simple tip for cedar identification is to look at the angle of the branches and think of the first letter of the tree's name: Deodar = Drooping; Lebanon = Level; Atlas = Ascending.

Deodar bark

Deodar needles

Deodar *Cedrus deodara* (Pinaceae) 36m
Broadly conical evergreen with drooping leading shoot on a tapering crown. BARK Almost black on old trees, fissured into small plates. BRANCHES With drooping tips. LEAVES In whorls of 15–20 on short lateral shoots, or in spirals on larger twigs. Needles 2–5cm long, shortest on lateral shoots, dark green with pale grey lines on either side. REPRODUCTIVE PARTS Male flowers purplish, turning yellow with autumn pollen release; to 12cm long. Mature female cones solid and barrel-shaped; to 14cm long and 8cm across, growing only on older trees. STATUS AND DISTRIBUTION Native of W Himalayas. Introduced to Britain in 1831 and widely planted in parks and gardens, where it can form a stately tree.

Atlas Cedar needles

Atlas Cedar bark

Atlas Cedar *Cedrus atlantica*
(Pinaceae) 40m
Broadly conical or pyramidal tree, domed when mature. Leading shoot usually rises above domed top. BARK Dark grey, cracking into large plates with deep fissures. BRANCHES Tips angled upwards. Shoots short and ascending. LEAVES Shiny, deep green, 1–3cm long, in clusters. REPRODUCTIVE PARTS Male cones 3–5cm long, pinkish yellow. Ripe female cones squat, with a sunken tip and small central boss; to 8cm long and 5cm across. STATUS AND DISTRIBUTION Native of Atlas Mountains of N Africa; widely planted for ornament. COMMENTS Most frequent form is Blue Atlas Cedar *C. atlantica* var. *glauca*, with bright bluish-grey foliage; hardy, tolerates atmospheric pollution.

Cedar of Lebanon needles

Cyprus Cedar needles

Cedar of Lebanon *Cedrus libani* (Pinaceae) 40m
Mature tree is flat-topped, with immense trunk in old trees. BARK Dark grey, fissured and ridged, becoming dark brown in very old trees. BRANCHES Main ones massive and ascending; smaller, lateral branches level, supporting flat plates of foliage. LEAVES Needles, to 3cm long, usually in clusters of only 10–15 on short shoots, singly if growing on long shoots. REPRODUCTIVE PARTS Male cones greyish or blue-green and erect, to 7.5cm long. Mature female cones solid, ovoid, to 12cm long and 7cm across, ripening from purple-green to brown. STATUS AND DISTRIBUTION Native of mountain forests of E Mediterranean; widely planted in British parks and gardens since 1640.

SIMILAR TREE
Cyprus Cedar *C. brevifolia* (21m) Needles dark green and shorter than those of other cedars (2cm), and crown more open. Female cones to 7cm long, ripening from purple-green to brown. Native of Troodos Mountains on Cyprus; sometimes grown in British collections.

Deodar flowers and foliage

Deodar cones and foliage

Deodar

Atlas Cedar cones and foliage

Blue Atlas Cedar cones and foliage

Blue Atlas Cedar

Cyprus Cedar foliage

Cedar of Lebanon foliage

Cedar of Lebanon

LARCHES *LARIX* (FAMILY PINACEAE)

A genus of 10 species, some very widespread around the northern tree-lines of Asia, Europe and North America. Larches are conifers, but all of them are deciduous, some producing fine autumn colours. The leaves are produced in short spirals on new shoots, and in whorls on tiny spurs produced on older wood. The cones are small, woody and persistent. Freshly cut foliage has a pleasant scent of new-mown hay. Many are commercially important timber species.

Common Larch *Larix decidua* (Pinaceae) 35m

A deciduous conifer, forming a tall, narrowly conical tree if growing alone, but more often seen in close rows in plantations. BARK Rough and greyish brown in young trees, becoming fissured with age. BRANCHES Mostly horizontal, but lower ones on old trees slightly drooping. LEAVES Needles in tight bunches of up to 40, each needle up to 3cm long, fresh green when first open, becoming darker, with 2 pale bands below in summer, and then changing through red to yellow before falling in autumn. REPRODUCTIVE PARTS Male flowers small, soft, yellow cones, releasing pollen in spring. Female cones conspicuously red in spring, maturing to become woody, brown and ovoid. Cones ripen in the first year but persist on the twigs after releasing their seeds. STATUS AND DISTRIBUTION A native of the mountains of central and E Europe but long established in Britain as a timber species, or as an ornamental tree in gardens. It tolerates calcareous soils. COMMENTS The seeds are a favourite food of the Common Crossbill, a bird whose bill is uniquely adapted to parting the cone scales to extract the seeds. Being deciduous, plantations of Common Larch are light and airy during winter and spring. Consequently, spring woodland flowers usually flourish to a far greater extent than under other conifers. The bark can be used for tanning, while the trees themselves make straight poles.

Common Larch foliage

Common Larch mature cone

Japanese Larch *Larix kaempferi* (Pinaceae) 40m

A deciduous conifer resembling Common Larch, but lacking the drooping shoots, and having a more twiggy appearance with a dense crown. BARK Reddish brown, flaking off in scales. LEAVES Needles, growing in tufts of about 40, slightly broader, and greyer in colour than those of Common Larch. REPRODUCTIVE PARTS Male cones similar to those of Common Larch, but female cones pink or cream in spring, becoming brown and woody in autumn, and differing from those of Common Larch in having turned-out tips to the scales, looking like woody rose-buds. STATUS AND DISTRIBUTION Native of Japan, but now very common in forestry plantations, replacing the Common Larch because of its more vigorous growth. Of less value to wildlife, because of the dense needle-litter that accumulates beneath it, and the later leaf-fall.

Japanese Larch foliage

Japanese Larch mature cone

Common Larch bark

Common Larch cones and foliage

Common Larch in winter

Common Larch

Japanese Larch bark

Japanese Larch autumn colours

Japanese Larch cones

Japanese Larch foliage

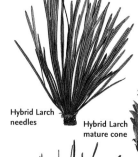

Hybrid Larch *Larix × marschlinsii* (Pinaceae) 32m

Vigorous deciduous conifer, conical in outline when mature. Shares characteristics with both parents (Common Larch and Japanese Larch; *see* p. 106); most features intermediate between the 2 but rather variable. BARK Similar to Japanese Larch. LEAVES Dark green needles, to 5cm long. REPRODUCTIVE PARTS Female cones pinkish at first, but ripening yellow-brown; scales slightly reflexed with projecting bracts. STATUS AND DISTRIBUTION More vigorous than either parent, and copes better with harsh conditions and poor soils; quite widely planted, mostly for timber but occasionally for ornament.

Hybrid Larch needles

Hybrid Larch mature cone

Tamarack *Larix laricina* (Pinaceae) 20m

A very slender, upright tree, the North American counter-part of Common Larch, with the smallest cones and flowers of any larch. BARK Pinkish and scaly. BRANCHES Twisted, with curled shoots. LEAVES Dark green, narrow needles with grey bands below. REPRODUCTIVE PARTS Cones purplish, to 2cm long with 15–20 scales. STATUS AND DISTRIBUTION Native to N North America. Planted in Britain occasionally for ornament.

Tamarack needles

Tamarack mature cone

Dahurian Larch *Larix gmelinii* (Pinaceae) 30m

Slender, conical deciduous tree. BARK Reddish brown and scaly. BRANCHES Level, sometimes forming flattish areas of foliage, and supporting long yellowish or red-brown downy shoots. LEAVES Blunt-tipped needles, bright green above with 2 paler bands below, to 4cm long; in clusters of 25. REPRODUCTIVE PARTS Female cones similar to other larches, with pinkish or greenish slightly projecting bracts, becoming brown when ripe, with square-ended scales. STATUS AND DISTRIBUTION Native of E Asia, sometimes planted in Britain for timber or as a specimen tree. COMMENTS Prince Rupert's Larch, with larger cones, is the most widely planted variant.

Dahurian Larch needles

Dahurian Larch mature cone

Western Larch *Larix occidentalis* (Pinaceae) 30m

Largest of all the larches, although it rarely reaches its maximum height away from its native range. A tall, slender, conical tree. BARK Grey and scaly, forming deep fissures low down. BRANCHES Slightly ascending and short with red-brown shoots. LEAVES Soft needles up to 4cm long, in tufts on side-shoots. REPRODUCTIVE PARTS Male flowers yellow, and pendent below the shoots; female flowers red and upright above the shoot on the same tree. Both open in spring. Cones ovoid, 4cm long, with long bracts protruding from between the scales, distinguishing this from all other larches. STATUS AND DISTRIBUTION Native of mountains of British Columbia, S to Oregon. Introduced in 1881 to Britain, where there are now some very fine specimens in mature collections.

Western Larch needles

Western Larch mature cone

Hybrid Larch cones and foliage

Tamarack cone and foliage

Tamarack bark

Dahurian Larch young cones and foliage

Western Larch

Prince Rupert's Larch cone and foliage

Western Larch cone and foliage

SPRUCES *PICEA* (FAMILY PINACEAE)

About 50 species occur in the northern hemisphere, found high on mountains in the south, but down to lower levels further north. All are pleasingly conical, with rough, scaly bark that does not form ridges, tiny pegs left on shoots after needles have fallen and tough, often spined leaves. Cones are long and pendulous when ripening, and usually fall from the tree intact. Many species are important timber trees and widely planted.

ABOVE: **Norway Spruce bark**
RIGHT: **Norway Spruce needles**
FAR RIGHT: **Norway Spruce cone**

Norway Spruce *Picea abies* (Pinaceae) 44m

Familiar evergreen, commonly used in its early years as the Christmas tree. Narrowly conical tree on a slender, unbranching trunk. BARK Brownish, scaly, with resinous patches on older trees. BRANCHES Almost level. LEAVES Stiff, short needles, 4-angled on short pegs, spreading to expose undersurface of twig. REPRODUCTIVE PARTS Male cones small, yellowish, clustered near tips of shoots; female cones up to 18cm long, narrowly oval and pendulous. STATUS AND DISTRIBUTION Native of European mountains, and at lower altitudes further north. Widely planted in Britain, especially as Christmas trees and shelter-belts.

ABOVE: **Blue Colorado Spruce bark**
RIGHT: **Blue Colorado Spruce needles**
FAR RIGHT: **Blue Colorado Spruce cone**

Blue Colorado Spruce *Picea pungens* (Pinaceae) 30m

Slender conical evergreen. BARK Purplish and ridged. BRANCHES With smooth, yellowish-brown twigs. LEAVES Sharply pointed, stiff needles, to 3cm long, usually dark green (in some cultivars markedly blue-green); growing all round shoot, but with more on upper surface, and some curving upwards to make top surface look more dense. REPRODUCTIVE PARTS Male and female flowers in small, separate clusters on same tree; males red-tinged, females greener. Mature female cones pendent, narrowly oval, to 12cm long, often slightly curved; scales have irregularly toothed tips. STATUS AND DISTRIBUTION A native of SW USA, growing on dry, stony mountain slopes and stream sides, but commonly planted for ornament and timber throughout much of N Europe. COMMENTS Most commonly seen cultivar is var. *glauca* (23m), favoured for its attractive bluish foliage.

ABOVE: **Brewer's Spruce bark**
RIGHT: **Brewer's Spruce needles**

Brewer's (Weeping) Spruce *Picea breweriana* (Pinaceae) 20m

Markedly conical evergreen with a slender bole. BARK Grey-purple, scaly. BRANCHES With pale brownish or pink, downy twigs. Note the striking 'weeping' habit of shoots along branches. LEAVES Flattened, needle-like, sharply pointed, to 3cm long, green above with white bands below; growing all round shoot and often curving forwards. REPRODUCTIVE PARTS Male flowers large for a spruce, to 2cm across, reddish; female cones pendent, cylindrical, to 12cm long, starting purplish but ripening brown. Overlapping scales have blunt, rounded tips. STATUS AND DISTRIBUTION Native to W USA. Popularly planted in the British Isles and graceful in maturity.

Norway Spruce cones and foliage

Norway Spruce

Blue Colorado Spruce foliage

Blue Colorado Spruce

Blue Colorado Spruce var. *glauca*

Brewer's Spruce foliage

Brewer's Spruce

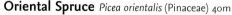

ABOVE: Oriental Spruce bark

RIGHT: Oriental Spruce needles

Oriental Spruce *Picea orientalis* (Pinaceae) 40m

Dense-foliaged evergreen growing into a strongly conical tree on a short, stout bole. BARK Pale brown, scaly. BRANCHES Slender, with numerous hairy twigs. LEAVES Very short, blunt needles, to 1cm long, arising all round shoots, but leaving more open area on the lower surface; dark green and glossy above, square in cross section. REPRODUCTIVE PARTS Small male flowers red then yellow. Female cones to 8cm long, pendent, ovoid, often curved and green with purple or grey tinges when still growing, ripening to shiny brown. STATUS AND DISTRIBUTION Native of mountain forests of Caucasus and NE Turkey, widely planted in the British Isles for ornament, and occasionally for commercial forestry. COMMENTS The variety 'Aurea', with bright yellow young foliage, is a popular arboretum tree.

White Spruce needles

White Spruce *Picea glauca* (Pinaceae) 24m

Narrowly conical evergreen, broadening with maturity. BARK Purple-grey with roughly circular scales. BRANCHES Turning upwards at tips, with hairless, greyish twigs and blunt buds. LEAVES Pointed needles, 4-angled, to 1.3 cm long, pale green (sometimes bluish), and smelling unpleasantly to some when crushed. REPRODUCTIVE PARTS Female cones about 6cm long and 2cm across, cylindrical, pendent, orange-brown when ripe, with rounded margins to scales. STATUS AND DISTRIBUTION Native of N North America, widely planted in the British Isles for timber and ornament.

ABOVE: Black Spruce bark

RIGHT: Black Spruce needles

Black Spruce *Picea mariana* (Pinaceae) 19m

Slender, conical evergreen with the shortest needles and cones of any spruce (apart from Oriental, whose needles are darker green and blunt). BARK Grey-brown and scaly. LEAVES Bluntly pointed needles, blue-green above and pale blue below, to 1.5cm long, 4-angled, growing all round hairy, yellowish shoots. REPRODUCTIVE PARTS Cones ovoid, reddish and pendent, to 4cm long, usually growing near tree top. STATUS AND DISTRIBUTION Native of North America, planted in the British Isles for ornament.

Engelmann's Spruce *Picea engelmannii* (Pinaceae) 30m

Slender, conical evergreen. Trunk thin and narrowly tapering. BARK Greyish pink, scaly. BRANCHES Ascending, turning upwards at tips, with pendulous young shoots. LEAVES Pointed, 4-angled, bluish-green needles, to 2.5cm long, spreading to reveal twig's lower surface but hiding upper surface; smelling unpleasant when crushed. REPRODUCTIVE PARTS Narrowly oval cones, tapering to a point, to 7cm long, ripening brownish, with squarish toothed scales. STATUS AND DISTRIBUTION Native of Rocky Mountains in North America. Planted in Britain, but scarce. COMMENTS Ornamental variety 'Glauca' resembles Colorado Blue Spruce with its bluish foliage, but leaves are softer, smelling pleasantly of menthol; bark is orange-brown with thin papery scales, and cones are small and shiny brown.

ABOVE: Engelmann's Spruce bark

RIGHT: Engelmann's Spruce needles

Oriental Spruce foliage

Oriental Spruce mature cone and foliage

Oriental Spruce

Black Spruce foliage

Engelmann's Spruce foliage

White Spruce

Engelmann's Spruce foliage

Sitka Spruce *Picea sitchensis* (Pinaceae) 52m

Sitka Spruce needles

Large conical evergreen tapering to a spire-like crown. Trunk stout and buttressed in large specimens. BARK Greyish brown, becoming purplish and scaly in older specimens. BRANCHES Ascending with slightly pendent, hairless side-shoots. LEAVES Needles, to 3cm long, stiff and flattened with a distinct keel, bright green above with 2 pale blue bands below; appearing crowded on upper surface of shoot, with lower surface more exposed. General impression is of tough, sharply spined, blue-green foliage on a sturdy tree. REPRODUCTIVE PARTS Female cones yellowish, small at first, growing to about 9cm, becoming cylindrical and shiny pale brown, covered with papery, toothed scales. STATUS AND DISTRIBUTION Native of high-rainfall areas on W coast of North America. The largest spruce species: some specimens, guarded in national parks, have reached heights of 80m. Introduced to Britain and widely planted for commercial forestry and sometimes for ornament. COMMENTS Fast-growing in good conditions, growing 1m a year when young, even in the poorest soils. So, popular for commercial forestry and often grown in vast plantations, providing lightweight, strong timber.

Serbian Spruce needles

Serbian Spruce *Picea omorika* (Pinaceae) 30m

Narrowly conical to columnar tree, with a slender form unlike all other spruces. BARK Orange-brown and scaly in older trees. BRANCHES Lower branches slightly descending with raised tips, higher branches mostly level or ascending. All branches short. LEAVES Flattened, keeled needles, to 2cm long, may be blunt or barely pointed, dark blue-green above with 2 pale bands below. REPRODUCTIVE PARTS Male cones large and red, becoming yellow when releasing pollen. Female cones, on curving stalks, up to 6cm long, ovoid and blue-green at first, ripening to brown. Cone scales rounded with finely toothed margins. STATUS AND DISTRIBUTION Native of limestone rocks of the Drina basin of Serbia, unknown until 1875, but now a popular ornamental tree because of its pleasing shape, its tolerance of a wide range of soil types, its ability to grow in polluted air near towns, and its resistance to frost damage.

Sargent's Spruce needles

Sargent's Spruce *Picea brachytyla* (Pinaceae) 26m

Broadly conical spruce with a tapering crown and a fairly open habit. BARK Grey-brown with squarish scales. BRANCHES Ascending, with pendulous side-shoots. LEAVES Crowded, pointed needles with a green upper surface and a pale to silvery white lower surface. REPRODUCTIVE PARTS Male flowers small and reddish yellow, female cones narrowly ovoid with triangular-tipped scales. STATUS AND DISTRIBUTION Introduced to Britain from China in the early 20th century, and now mostly confined to gardens and collections.

Sitka Spruce bark

Sitka Spruce foliage

Sitka Spruce

Serbian Spruce

Serbian Spruce cone and foliage

Serbian Spruce foliage

Sargent's Spruce flowers and foliage

Sargent's Spruce cone and foliage

HEMLOCK-SPRUCES *TSUGA* (FAMILY PINACEAE)

About 10 species, related to spruces, but lacking small pegs on shoots. Cones small, woody and usually pendent. Timber used for paper-pulp. Gets its English name because leaves supposedly smell of the poisonous herb Hemlock.

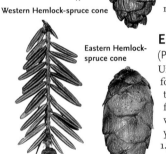

Western Hemlock-spruce *Tsuga heterophylla*

(Pinaceae) 45m

Large, narrowly conical evergreen with dense foliage; crown spire-like with drooping leading shoot. LEAVES Needles, dark glossy green above with 2 pale bands below, in 2 flattened rows on either side of shoot. Note 2 leaf sizes (hence *heterophylla*): some 6mm long, others to 2cm long, both with rounded tips and toothed margins. REPRODUCTIVE PARTS Male flowers reddish at first, but yellowing with pollen. Female cones solitary, ovoid and pendent, to 3cm long; scales blunt. STATUS AND DISTRIBUTION Native of W North America. Widely planted in the British Isles, and reaches a great size. Grows well on most soils except chalk; prefers high-rainfall areas.

Western Hemlock-spruce needles

Western Hemlock-spruce cone

Eastern Hemlock-spruce cone

Eastern Hemlock-spruce *Tsuga canadensis*

(Pinaceae) 30m

Untidy tree with heavy branches, a forked trunk and dark foliage. BARK Blackish. LEAVES Needles, more tapering than those of Western Hemlock-spruce and with a narrower tip; further row of leaves along middle of shoot twists to show white undersides. REPRODUCTIVE PARTS Male flowers small, yellowish, clustered along underside of shoots. Female cones 1.5cm long; cone scales have thickened edges. STATUS AND DISTRIBUTION Native of E North America. Widely planted in the British Isles.

Eastern Hemlock-spruce needles

DOUGLAS FIRS *PSEUDOTSUGA* (FAMILY PINACEAE)

Scientific name means 'false hemlock'; foliage is soft and bark deeply fissured in large trees. Cones, which have 3-pointed bracts projecting between scales, are pendent and shed whole when ripe.

Douglas Fir *Pseudotsuga menziesii* (Pinaceae) 60m

Tall, slender, conical evergreen. BARK Greyish green, smooth with resinous blisters in young trees; turns red-brown with age. BRANCHES In whorls supporting pendulous masses of dense aromatic foliage. LEAVES Needles, to 3.5cm long, blunt or slightly pointed, dark green and grooved above, with 2 white bands below. REPRODUCTIVE PARTS Male flowers small, yellow and pendulous; growing near tips of twigs. Female flowers, like tiny pinkish shaving-brushes, grow at tips of twigs; ovoid when ripe, pendulous and brown with unique 3-tailed bracts between scales. STATUS AND DISTRIBUTION Native of

ABOVE: Douglas Fir bark
RIGHT: Douglas Fir needles

W North America, where it reaches an immense size. Planted in the British Isles for timber and ornament, doing well in Scotland. COMMENTS Named after Scottish explorer David Douglas. Annual rings easy to count as wood put down in summer is very dark.

Western Hemlock-spruce bark

Western Hemlock-spruce cones and foliage

Western Hemlock-spruce

Eastern Hemlock-spruce cones and foliage

Eastern Hemlock-spruce cone

Douglas Fir trunk

Douglas Fir male flowers and foliage

Douglas Fir cone and foliage

PINES *PINUS* (FAMILY PINACEAE)

The largest genus of conifers, with over 100 species occurring in most climatic regions of the northern hemisphere. They are commercially very important, many of them providing good softwood timber. Cones are generally large, woody and pendent, and the long needles grow in bunches of 2, 3 or 5. Most of the northern species are 2-needle pines.

Scots Pine *Pinus sylvestris* (Pinaceae) 36m

A conical evergreen when young and growing vigorously, but becoming much more open, and flat-topped with a long bole, when an older tree. BARK Red- or grey-brown low down on the trunk, but markedly red or orange higher up in mature trees. Lower trunk scaly, but more papery higher up. BRANCHES Irregular, with broken-off stumps of old branches remaining on the trunk lower down. LEAVES Needles, in bunches of 2, grey-green or blue-green, up to 7cm long, usually twisted with a short point at the tip. REPRODUCTIVE PARTS Male flowers yellow, in clusters at ends of previous year's shoots, shedding pollen in late spring. Female flowers at tips of new shoots; usually solitary, crimson at first, ripening to brown by late summer and persisting through winter. In the second summer they enlarge and become green and bluntly conical, ripening to grey-brown in autumn; they do not open their scales and shed seeds until the following spring. Each cone scale has a blunt projection in the centre. STATUS AND DISTRIBUTION Native to Scotland, and originally much of Britain, as well as a wide swathe of Europe and Asia from Spain to Siberia and Turkey. Introduced to the USA, where it is often used as a Christmas tree. COMMENTS A useful timber tree, producing 'deal', used for pit-props, telegraph poles, building, furniture, chipboard and paper-pulp. Native pinewoods are confined to Scotland, referred to there as Caledonian forests, and supporting a unique assemblage of wildlife, including species that cannot survive in plantations of alien species. In particular, the Scottish Crossbill is found nowhere else in the world, its bill adapted to extract Scots Pine seeds from the cones. Other species associated with these native forests include Crested Tit, Capercaillie, Red Squirrel and Pine Marten. Elsewhere in Britain and Ireland, the Scots Pine has been planted. It often colonises heathland and in such circumstances is usually viewed as an unwelcome invader, open heathland being such a threatened and restricted habitat in the region.

Scots Pine needles

Scots Pine cone

Scots Pine cone nibbled by Red Squirrel

Capercaillie, a Caledonian forest specialist

Scots Pine bark

'Caledonian Pine'

Scots Pine cones and foliage

Scots Pine cones and foliage

Pine Hawk-moth larva

Scots Pine, naturalised tree

Japanese Red Pine *Pinus densiflora* (Pinaceae) 15m

Young trees are the most attractive, with a neat conical shape. Older trees are less shapely and have a flatter, twiggier crown. BARK Distinctly red and flaky. BRANCHES Usually drooping, with bright green shoots showing clearly among the rather sparse foliage. LEAVES Slender, 8–12cm long and shiny green. REPRODUCTIVE PARTS Cones pointed and reddish, to 5cm long. Flowers and cones are produced on quite young trees. STATUS AND DISTRIBUTION Native of China, Japan and Korea, occasionally planted in Britain.

Red Pine *Pinus resinosa* (Pinaceae) 20m

Straggly evergreen, very similar to Scots Pine. BARK Reddish and scaly. BRANCHES Usually drooping. LEAVES 10–15cm long, longer and more slender than those of Scots Pine and widely spaced, with the unique characteristic of snapping easily when bent in a curve. REPRODUCTIVE PARTS Cones reddish brown, ovoid. STATUS AND DISTRIBUTION Native of Great Lakes area of USA and Canada, occasionally planted in Britain.

Japanese Black Pine *Pinus thunbergii* (Pinaceae) 25m

Often rather wizened-looking with a rather sparse appearance. BARK Very dark and deeply fissured in older trees. BRANCHES Straggling, lower branches drooping. LEAVES Rigid, spined, to 12cm long, in whorls. REPRODUCTIVE PARTS Cones ovoid, brown, often in dense clusters. STATUS AND DISTRIBUTION Native of Japan and Korea, occasionally planted in Britain.

Red Pine needles

Beach and Lodgepole Pines *Pinus contorta* (Pinaceae) 30m

Small to medium-sized evergreen that occurs as 2 different subspecies. Beach Pine (ssp. *contorta*) grows near the shore from Alaska to California; scientific name gained from twisted appearance of branches contorted by wind. Needles paired, up to 7cm long, with sharp points; usually twisted and densely packed on young shoots, but sparser on older shoots. Lodgepole Pine (ssp. *latifolia*) is very similar, but more columnar with a less dense crown; usually grows on a much straighter but sometimes divided trunk. Needles broader than those of Beach Pine, and more spread apart. Grows in the mountains well inland away from the sea. BARK Blackish brown in all trees. REPRODUCTIVE PARTS All trees have male flowers in dense clusters near tips of shoots. Female flowers in groups of up to 4 close to tip of shoot. Cones rounded to ovoid, up to 6cm long and 3cm across, and usually a shiny yellow-brown. Each cone scale has a slender, sharp tip, which easily breaks off. STATUS AND DISTRIBUTION Native to coastal W North America, widely planted for timber in the British Isles on poor soils and exposed, often upland sites.

Lodgepole Pine flowers and foliage

Japanese Red Pine bark

Japanese Red Pine foliage

Red Pine bark

Red Pine cones and foliage

Japanese Black Pine foliage

Japanese Black Pine

Lodgepole Pines

Bosnian Pine *Pinus heldriechii (leucodermis)* (Pinaceae) 30m

Broadly pyramidal tree with a tapering bole. BARK Grey, with irregular plates. Whitish patches appearing with age. LEAVES Paired needles, to 9cm long, densely packed on shoots, stiff and projecting at right angles, pungent. REPRODUCTIVE PARTS Cones, to 8cm long and 2.5cm across, narrowly ovoid and ripening to brown; scales have a recurved prickle. Second-year cones deep blue. STATUS AND DISTRIBUTION Native of Balkans and SW Italy, mainly on dry mountain limestone. Planted elsewhere for ornament, thriving on free-draining soils.

Bosnian Pine needles

Austrian Pine cone

Austrian Pine *Pinus nigra* ssp. *nigra* (Pinaceae) 30m

Broadly conical, heavily branched tree; usually has single bole and narrow crown. BARK Greyish brown, rough in older trees. LEAVES Paired needles, to 15cm long, flattened and stiff with finely toothed margins; lasting for up to 4 years, creating dense foliage. REPRODUCTIVE PARTS Mature cones, to 8cm long, have keeled, spined scales; solitary or in small clusters. STATUS AND DISTRIBUTION Native of mountains of central Europe; variable across range. Planted in the British Isles for shelter or ornament; tolerates a range of soils and climates.

Austrian Pine needles

Austrian Pine bark

Corsican Pine *Pinus nigra* ssp. *maritima* (Pinaceae) 30m

Similar to ssp. *nigra* but more shapely. BRANCHES Shorter than ssp. *nigra* and level, so young trees are columnar. LEAVES Soft, narrow needles, paler green than ssp. *nigra*, to 15cm long, often twisted in young trees. REPRODUCTIVE PARTS Cones similar to ssp. *nigra*. STATUS AND DISTRIBUTION Native of Corsica, S Italy and Sicily, planted in the British Isles on lowland heaths, coastal dunes, and poor soils. Resistant to pollution.

Corsican Pine needles

Crimean Pine *Pinus nigra* var. *carmanica* (Pinaceae) 30m

Distinguished by strong bole that divides into several vertical stems, growing upright close to each other. Native to Crimea and Asia Minor. Planted in British parks and gardens.

Corsican Pine bark

Mountain Pine *Pinus mugo* (Pinaceae) 30m

Two forms: tree-sized ssp. *uncinata* and shrub-like ssp. *mugo*. BARK Greyish black in all trees. LEAVES Bright green needles in all trees, to 8cm long, curved and stiff, appearing whorled. REPRODUCTIVE PARTS All trees have male flowers in clusters near shoot tips; female flowers reddish, in groups of 1–3. Ripe cones ovoid, pale brown, to 5cm long; scales have a small prickle. STATUS AND DISTRIBUTION Native of Alps, Pyrenees and Balkans; dwarf forms occur at high altitudes. The ssp. *uncinata* is planted for forestry, shelter, stabilising sand etc. Dwarf forms are favoured for ornament.

Mountain Pine bark

Bosnian Pine cones and foliage

Bosnian Pine flowers and foliage

Austrian Pine

Crimean Pine

Austrian Pine cone and foliage

Corsican Pine cones and foliage

Mountain Pine foliage

Mountain Pine

Stone Pine *Pinus pinea* (Pinaceae) 30m

Broad umbrella-shaped tree with a dense mass of foliage on spreading branches on top of a tall bole. BARK Reddish grey on old trees and fissured, flaking away to leave deep orange patches. LEAVES Paired needles, to 20cm long and 2mm wide, slightly twisted. Through a hand-lens 12 lines of stomata can be seen on outer surface and 6 on inner surface. REPRODUCTIVE PARTS Cones rounded to ovoid, to 14cm long and 10cm across, ripening rich glossy brown after 3 years. Scales, closely packed, with a slightly pyramidal surface, conceal large, slightly winged seeds. STATUS AND DISTRIBUTION Native of Mediterranean coasts. Planted in the British Isles occasionally, usually near coasts. COMMENTS Large, edible seeds are harvested and sold as 'pine kernels' or 'pine nuts'.

Stone Pine needles

Maritime Pine *Pinus pinaster* (Pinaceae) 32m

Has a sturdy, slightly tapering bole, often curved in exposed coastal areas, the crown fairly open, reflecting curve of bole. BARK Yellowish brown, breaking into rectangular flakes. LEAVES Needles, the longest and thickest of any 2-needle pine. REPRODUCTIVE PARTS Male flowers yellow and ovoid, in clusters near shoot tips. Female cones ovoid, red at first, in small clusters, ripening conical and woody with a greenish-brown gloss. STATUS AND DISTRIBUTION Native of SW Atlantic coasts of Europe and Mediterranean. Grows well on poor sandy soils, often on heaths and near coasts.

Maritime Pine needles

Aleppo Pine *Pinus halepensis* (Pinaceae) 20m

A small pine, often growing in a gnarled and deformed manner, but sometimes maturing to form a broad, shapeless crown on a stout bole. BARK Shiny, smooth, silver-grey in young trees, becoming scaly, fissured and redder with age. BRANCHES Twigs characteristically pale grey, or even white. LEAVES Paired needles slender (0.7mm), to 15cm long, sometimes slightly twisted and with very finely toothed margins. REPRODUCTIVE PARTS Red-brown cones up to 12cm long and 4cm across, oval or conical and borne singly on short stalks, or in groups of 2–3, and sometimes deflexed. Cone scales shiny reddish brown, hiding winged seeds up to 2cm long. STATUS AND DISTRIBUTION Widespread and common around the Mediterranean, planted in the British Isles for ornament.

Calabrian Pine needles

Calabrian Pine *Pinus brutia* (Pinaceae) 20m

Closely related to Aleppo Pine, but leaves broader (1–1.5mm), darker green and stiffer. Twigs reddish yellow or greenish. Cones spreading out from the twig and never deflexed. Occurring in similar places, but further east, in Calabria, Crete, Cyprus and Turkey, where it can form open forests on coastal hills. Planted occasionally in the British Isles.

Stone Pine bark

Stone Pine flowers and foliage

Maritime Pine cones and foliage

Aleppo Pine cone and foliage

Aleppo Pine foliage

Aleppo Pine foliage

Aleppo Pine bark

Calabrian Pine bark

Calabrian Pine foliage

Monterey Pine bark

Monterey Pine *Pinus radiata* (Pinaceae) 45m

Large, variable pine, slender and conical when growing vigorously, becoming more domed and flat-topped on a long bole with age. BARK Fissured and grey, blackening with age. BRANCHES Main ones sometimes hanging low enough to touch ground. LEAVES Bright green needles in bunches of 3; each needle thin and straight, to 15cm long, with a finely toothed margin and sharp-pointed tip. REPRODUCTIVE PARTS Male flowers in dense clusters near ends of twigs, releasing pollen in spring. Female cones in clusters of 3–5 around tips of shoots, ripening to large, solid woody cones, to 15cm long and 9cm across, with a characteristic asymmetrical shape. Cone scales thick and woody, with rounded outer edges, concealing black, winged seeds. STATUS AND DISTRIBUTION Native to a small area around Monterey, California, Guadalupe Island and Baja California, Mexico. Widely planted in mild areas of the British Isles as a shelter-belt tree or for ornament, growing well next to the sea. Vigorous when young. COMMENTS Adapted to live in areas subject to bush fires. Cones open to release seeds only when heated by fire; seeds germinate and grow rapidly in the ash, giving young trees a head start after a severe bush fire. Cones can be encouraged to release seeds by heating them in an oven, and seedlings are easy to grow.

Monterey Pine needles

Northern Pitch Pine needles

SIMILAR TREE

Northern Pitch Pine *P. rigida* (20m) Narrowly conical tree with stiff, tough needles in clusters of 3, and small cylindrical or rounded cones with thinner, but stiff (hence *rigida*) scales. A striking feature, unique to this species of pine, is sprouting foliage on the bole. Native to E coasts of North America, occasionally planted in Britain.

Ponderosa Pine (Western Yellow Pine)

Pinus ponderosa (Pinaceae) 40m (50m)

ABOVE: Ponderosa Pine bark
LEFT: Ponderosa Pine needles

Large, slender, conical pine with a sturdy, straight bole. BARK Scaly, pinkish brown. LEAVES Needles, to 30cm long, narrow (3mm) and stiffly curved with finely toothed edges and a sharp, pointed tip; clustered densely on shoots and persisting for 3 years. REPRODUCTIVE PARTS Cones ovoid, up to 15cm long and 5cm across, on short stalks or directly on twigs, sometimes leaving a few scales behind when they fall; solitary or in small clusters. Cone scales oblong with swollen, exposed, ridged tips hiding 5cm-long, oval, winged seeds. STATUS AND DISTRIBUTION Native to W USA, planted in Britain mostly for ornament.

Jeffrey Pine needles

SIMILAR TREE

Jeffrey Pine *P. jeffreyi* (40m) The upland counterpart of Ponderosa Pine in its native W USA. Leaves bluer and cones larger, up to 30cm long, with scales bearing slender, curved spines. Bark blacker. Grows well in Britain, planted mainly for ornament.

Monterey Pine cones and foliage

Monterey Pine

Monterey Pine foliage

Northern Pitch Pine foliage

Northern Pitch Pine

Ponderosa Pine foliage

Jeffrey Pine foliage

Jeffrey Pine foliage

ABOVE: **Arolla Pine bark**
RIGHT: **Arolla Pine needles**

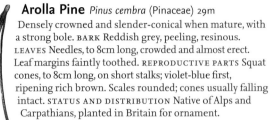

Arolla Pine *Pinus cembra* (Pinaceae) 29m
Densely crowned and slender-conical when mature, with a strong bole. BARK Reddish grey, peeling, resinous. LEAVES Needles, to 8cm long, crowded and almost erect. Leaf margins faintly toothed. REPRODUCTIVE PARTS Squat cones, to 8cm long, on short stalks; violet-blue first, ripening rich brown. Scales rounded; cones usually falling intact. STATUS AND DISTRIBUTION Native of Alps and Carpathians, planted in Britain for ornament.

Bhutan Pine *Pinus wallichiana* (Pinaceae) 35m
Narrowly columnar, becoming shapeless with age. BARK Greyish brown, resinous. BRANCHES Lower ones spreading, upper ones ascending. LEAVES Needles, to 20cm long and 7mm wide, supple with finely toothed margin. REPRODUCTIVE PARTS Cones long (to 25cm), cylindrical, growing below shoot, light brown and resinous. Cone scales wedge-shaped and grooved, thickened at tip. Basal scales sometimes reflexed. STATUS AND DISTRIBUTION Native of Himalayas, planted in Britain for ornament.

Bhutan Pine bark

Japanese White Pine needles

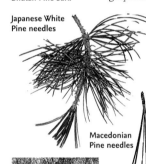

Macedonian Pine needles

Weymouth Pine *Pinus strobus* (Pinaceae) 32m
Mature tree has tapering trunk and rounded crown. BARK Dark grey. BRANCHES Level. LEAVES Blue-green needles; note tuft of hairs below each 10cm-long bunch of 5 needles. REPRODUCTIVE PARTS Slender cones; basal scales often curved outwards. STATUS AND DISTRIBUTION Native of North America, planted in Britain mainly for timber. COMMENTS Excellent timber quality, good for carving. The variety 'Contorta' has twisted needles and smaller habit.

SIMILAR TREE
Japanese White Pine *P. parviflora* (20m) Needles twisted, to 6cm long, blue-green outside and blue-white inside. Cones ovoid, to 7cm long, with tough scales. Native to Japan, planted in Britain occasionally.

Macedonian Pine bark

Macedonian Pine *Pinus peuce* (Pinaceae) 30m
Narrowly conical; trunk slender and crown pointed. BARK Greyish green. LEAVES Slender, supple needles, to 12cm long, with toothed margins and pointed tip. REPRODUCTIVE PARTS Cones to 20cm long, mostly cylindrical, sometimes curved near tip; growing below shoots, green, ripening to brown. STATUS AND DISTRIBUTION Native to Balkans, planted in Britain occasionally.

Rocky Mountain Bristlecone Pine bark

Rocky Mountain Bristlecone Pine *Pinus aristata* (Pinaceae) 10m
Small, slow-growing tree. LEAVES Needles 2–4cm long, dark green, often flecked with white resin; turpentine-scented and persisting for many years. REPRODUCTIVE PARTS Cones, to 6cm long; 6mm-long spine on each scale. STATUS AND DISTRIBUTION Native to Rocky Mountains, rarely planted in Britain. Bristlecone Pine *P. longaeva* has unspotted needles. An ancient tree in its native W USA; rare in Britain.

Arolla Pine foliage

Bhutan Pine foliage

Bhutan Pine cones and foliage

Macedonian Pine
INSET: cones and foliage

Weymouth Pine cones and foliage

Rocky Mountain Bristlecone Pine cone and foliage

WILLOWS *SALIX* (FAMILY SALICACEAE)

About 300 species worldwide, plus many hybrids. Most are shrubby, but a few form large trees.

Bay Willow leaf

Bay Willow *Salix pentandra* (Salicaceae) 18m

Broadly domed, open-crowned tree when growing in its typical open streamside habitat, but more slender and upright in woodlands. SHOOTS Olive-green and glossy. LEAVES Glossy green, showing a bluish tint beneath. REPRODUCTIVE PARTS In contrast to other willows, the bright yellow, upright male catkins appear at same time as new growth of leaves, rather than just before them. Female catkins dull yellowish green, longer and more pendulous. STATUS AND DISTRIBUTION A fairly common native of moorlands, streamsides and boggy areas, and also in damp, upland woods in north of region. Only very occasionally planted as an ornamental tree.

Crack-willow *Salix fragilis* (Salicaceae) 25m

Large tree when mature, with a broadly domed crown and a thick bole with a large base. BARK Dull grey-brown covered with thick interlocking criss-crossed ridges. BRANCHES Arising from low down near the base. SHOOTS Dull reddish brown, becoming brighter in early spring as leaves emerge. LEAVES Long and glossy, with toothed margins, widely spaced on the shoots. Lower surface less glossy and slightly paler than upper surface, and leaves have short green petioles. REPRODUCTIVE PARTS Male catkins yellow, pendulous, opening at about same time as leaves in early spring. Female catkins green, pendulous, on separate trees. STATUS AND DISTRIBUTION Very widespread native species, found in damp lowland woodlands and along river and canal banks. Hybridises freely, and not always true to the type species. COMMENTS Twigs snap easily and cleanly (hence *fragilis*) and root very readily if stuck in the ground. Even pieces that snap off accidentally and are carried along in a river will root if they lodge in a river bank. Often, lines of trees of the same sex grow along river banks, all derived from pieces of the same tree.

ABOVE: Crack-willow bark
RIGHT: Crack-willow leaf

White Willow leaf

Cricket-bat Willow leaf

White Willow *Salix alba* (Salicaceae) 25m

Large, broadly columnar tree with a dense crown when growing in the open, but more slender when crowded. BARK Dark grey. SHOOTS Yellowish grey, downy at first. LEAVES Bluish grey, smaller than similar Crack-willow. REPRODUCTIVE PARTS Male catkins small, elongated-ovoid, pendulous. Female catkins longer, slender and green. STATUS AND DISTRIBUTION Widespread native tree, often growing in damp lowland habitats. COMMENTS Easily raised from cuttings; grows well in damp soil, especially near ponds and rivers. Cricket-bat Willow, var. *caerulea*, has a straight trunk, purplish-red shoots and almost hairless leaves, bluish below. Planted as source of timber for cricket bats. So-called Coral-bark Willow cultivars are popular in parks and gardens.

Bay Willow foliage

Crack-willow

Crack-willow flowers and foliage

White Willow flowers and foliage

White Willow

Cricket-bat Willow foliage

Cricket-bat Willow

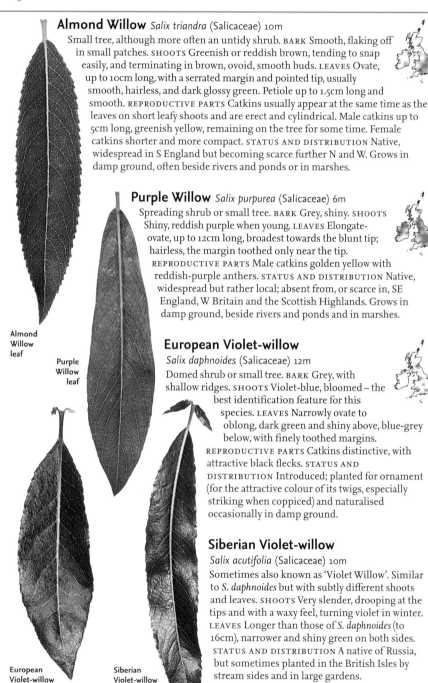

Almond Willow *Salix triandra* (Salicaceae) 10m

Small tree, although more often an untidy shrub. BARK Smooth, flaking off in small patches. SHOOTS Greenish or reddish brown, tending to snap easily, and terminating in brown, ovoid, smooth buds. LEAVES Ovate, up to 10cm long, with a serrated margin and pointed tip, usually smooth, hairless, and dark glossy green. Petiole up to 1.5cm long and smooth. REPRODUCTIVE PARTS Catkins usually appear at the same time as the leaves on short leafy shoots and are erect and cylindrical. Male catkins up to 5cm long, greenish yellow, remaining on the tree for some time. Female catkins shorter and more compact. STATUS AND DISTRIBUTION Native, widespread in S England but becoming scarce further N and W. Grows in damp ground, often beside rivers and ponds or in marshes.

Purple Willow *Salix purpurea* (Salicaceae) 6m

Spreading shrub or small tree. BARK Grey, shiny. SHOOTS Shiny, reddish purple when young. LEAVES Elongate-ovate, up to 12cm long, broadest towards the blunt tip; hairless, the margin toothed only near the tip. REPRODUCTIVE PARTS Male catkins golden yellow with reddish-purple anthers. STATUS AND DISTRIBUTION Native, widespread but rather local; absent from, or scarce in, SE England, W Britain and the Scottish Highlands. Grows in damp ground, beside rivers and ponds and in marshes.

European Violet-willow

Salix daphnoides (Salicaceae) 12m

Domed shrub or small tree. BARK Grey, with shallow ridges. SHOOTS Violet-blue, bloomed – the best identification feature for this species. LEAVES Narrowly ovate to oblong, dark green and shiny above, blue-grey below, with finely toothed margins. REPRODUCTIVE PARTS Catkins distinctive, with attractive black flecks. STATUS AND DISTRIBUTION Introduced; planted for ornament (for the attractive colour of its twigs, especially striking when coppiced) and naturalised occasionally in damp ground.

Siberian Violet-willow

Salix acutifolia (Salicaceae) 10m

Sometimes also known as 'Violet Willow'. Similar to *S. daphnoides* but with subtly different shoots and leaves. SHOOTS Very slender, drooping at the tips and with a waxy feel, turning violet in winter. LEAVES Longer than those of *S. daphnoides* (to 16cm), narrower and shiny green on both sides. STATUS AND DISTRIBUTION A native of Russia, but sometimes planted in the British Isles by stream sides and in large gardens.

Almond Willow leaf

Purple Willow leaf

European Violet-willow leaf

Siberian Violet-willow leaf

Almond Willow foliage

European Violet-willow catkins and winter twigs

Purple Willow foliage

Siberian Violet-willow foliage

European Violet-willow foliage

Chinese Weeping Willow *Salix babylonica* (Salicaceae) 20m

Has graceful 'weeping' branches and foliage reaching the ground. SHOOTS Brown and slender at first, becoming gnarled and thicker with age. LEAVES To 16cm long and 1.5cm wide, finely toothed and pointed; petiole to 5mm long. Mature leaves dark green, slightly glossy above. REPRODUCTIVE PARTS Catkins, to 2cm long and 0.4cm wide, appearing in May. STATUS AND DISTRIBUTION Native of China, planted in Britain and sometimes naturalised in wet habitats.

SIMILAR TREE
Weeping Willow *S.* × *sepulcralis* (20m) More popular than Chinese Weeping Willow. A hybrid between Chinese Weeping Willow and White Willow. Pendulous branches and golden foliage elegant in waterside settings.

Corkscrew Willow *Salix babylonica* 'Tortuosa' (Salicaceae) 18m

Distinctive willow, recognised by its contorted stems and pointed, twisted leaves. SHOOTS Older shoots less twisted, but bark still shows signs of earlier curves and even boles of older trees show some torsion. LEAVES Bright green, opening early in spring and darkening by summer. REPRODUCTIVE PARTS Male flowers are yellow catkins, to 2cm long. Female flowers smaller, greenish catkins, on separate plants. Both open around same time as leaves. STATUS AND DISTRIBUTION Originally from China but now found only in cultivation.

Chinese
Weeping Willow
leaf

Weeping
Willow
leaf

Corkscrew
Willow
leaf

Osier *Salix viminalis* (Salicaceae) 6m

Spreading shrub or small tree. Rarely reaches full potential, being regularly cropped for long flexible twigs ('withies'), used for weaving. Natural crown is narrow with slightly pendulous branches. SHOOTS Straight, flexible twigs, covered with greyish hairs when young, becoming smoother and shiny olive-brown with age. LEAVES Narrow, tapering, to 15cm long, the margin usually waved and rolled under; underside has grey woolly hairs. REPRODUCTIVE PARTS Male and female catkins, to 3cm long, appear before leaves on separate trees; erect or slightly curved. Males yellow, females browner. STATUS AND DISTRIBUTION Native in wet habitats but planted for withies, masking its native range.

Olive Willow *Salix elaeagnos* (Salicaceae) 6m

Similar to Osier. SHOOTS Young twigs with dense grey or white hairs; older twigs becoming yellow-brown and smooth. LEAVES Species is best recognised by leaves: *matt* white hairs beneath; dark shiny green above when mature. Leaves, to 15cm long and less than 1cm wide, have untoothed margins. REPRODUCTIVE PARTS Male and female catkins on separate trees, appearing just before leaves; reddish male catkins to 3cm long, female catkins smaller. STATUS AND DISTRIBUTION Native of mainland Europe, sometimes planted for ornament in the British Isles.

Osier leaf   Olive Willow leaf upperside
(*left*), underside (*right*)

Weeping Willow foliage

Corkscrew Willow foliage

Corkscrew Willow bark

Weeping Willow

Osier, pollarded trees

Osier bark

Osier foliage

Olive Willow foliage

Grey Willow *ssp. oleifolia* leaf upperside

Grey Willow *ssp. oleifolia* leaf underside

Grey Willow *ssp. cinerea* leaf underside

Grey Willow *Salix cinerea* (Salicaceae) 6m

Variable, usually a large shrub or sometimes a small tree with characteristic thick, downy, grey twigs. Represented by ssp. *cinerea* and ssp. *oleifolia* (previously known as *S. c. atrocinerea* or *S. atrocinerea*). SHOOTS Wood shows a series of fine longitudinal ridges if bark is peeled off 2-year-old twigs. LEAVES Oblong and pointed, usually 3–4 times as long as broad, on short petioles with irregular stipules. Leaves often have inrolled margins and are grey and downy below in spp. *cinerea*; by autumn, develop rusty hairs on veins in ssp. *oleifolia*. Upper surface matt and downy in ssp. *cinerea* but glossy and hairless in ssp. *oleifolia*. REPRODUCTIVE PARTS Catkins in early spring on separate trees, usually before leaves: male catkins ovoid and yellow, female catkins similar but greener, eventually releasing finely plumed seeds. This species and Goat Willow *S. caprea* are often called 'Pussy Willow' when their silky grey buds, resembling cats' paws, appear in spring, followed by bright yellow catkins. STATUS AND DISTRIBUTION Common across much of Britain, usually growing in wet habitats such as fenlands, stream sides and damp woodlands; ssp. *cinerea* is restricted mainly to East Anglia and Lincolnshire while ssp. *oleifolia* is widespread elsewhere.

Eared Willow *Salix aurita* (Salicaceae) 2m

Shrubby and much-branched willow. SHOOTS Downy at first, becoming shiny and brown with age. LEAVES Broadly ovate, to 4cm long, with wavy margins and a twisted tip; note the large, leafy stipules ('ears') at the leaf base. REPRODUCTIVE PARTS Male catkins ovoid and yellow, female catkins greener. STATUS AND DISTRIBUTION Favours damp, acid soils; common beside moorland and upland streams, and on damp heaths.

Eared Willow leaf

Dark-leaved Willow leaf

Tea-leaved Willow leaf

Dark-leaved Willow *Salix myrsinifolia* (previously known as *S. nigricans*) (Salicaceae) 4m

Branched, shrubby willow. SHOOTS Downy at first, becoming smoother and dull brown with age. LEAVES Ovate, to 7cm long, dark green above but glaucous below; note toothed margin and large stipules. Leaves turn black when dried. REPRODUCTIVE PARTS Males catkins ovoid, yellow; female catkins greener. STATUS AND DISTRIBUTION Native, favouring damp, stony and rocky ground. A northern and upland species.

Tea-leaved Willow *Salix phylicifolia* (Salicaceae) 4m

Much-branched shrub or small tree. SHOOTS Downy when young but smoother with age, becoming shiny reddish brown. LEAVES Ovate, to 8cm long, hairless with toothed margins; leathery texture, shiny green above but greyish below. REPRODUCTIVE PARTS Male catkins ovoid and yellow, female catkins greener. STATUS AND DISTRIBUTION Native, on damp, rocky ground.

Grey Willow bark

Grey Willow
ssp. *oleifolia* foliage

Eared Willow foliage

Eared Willow
catkins and foliage

Dark-leaved Willow foliage

Tea-leaved Willow foliage

Goat Willow leaf

Goat Willow (Sallow) *Salix caprea* (Salicaceae) 12m

Depending on its situation this may be a multi-branched, dense, shrubby tree, or a taller tree with a straight, ridged stem and sparsely domed crown. SHOOTS Thick, stiff twigs hairy at first, but becoming smoother and yellowish brown with age. Twigs smooth when bark is peeled off (compare with Grey Willow). LEAVES Large, up to 12cm long, oval, with a short twisted point at the tip. Upper surface dull green and slightly hairy, lower surface noticeably grey and woolly; in windy weather this can suddenly change the appearance of the tree from green to grey as the leaves are blown around. Leaf margins have small, irregular teeth, and short petiole sometimes has 2 ear-like sinuous stipules at its base. REPRODUCTIVE PARTS Male and female catkins, on separate trees, appear before the leaves, often in very early spring in sheltered places. Up to 2.5cm long, they are ovoid and covered with greyish silky hairs before opening; at this time, Goat Willow is often called 'Pussy Willow' because the silky grey buds bear a fanciful resemblance to cats' paws. Male catkins become bright yellow when they open; female catkins greener, producing numerous silky-haired seeds. STATUS AND DISTRIBUTION A widespread and common native species in Britain and Ireland, occurring in woods, hedgerows and scrub, and often in drier places than other similar species. COMMENTS An extremely important species in ecological terms, serving as a food plant for the larval or adult stages of a large number of insects. It is particularly noted for the number of species of Lepidoptera that are associated with it. In the case of most moth species, the larvae consume the leaves: the Sallow Kitten is a good example. However, the larvae of the Sallow Clearwing and Lunar Hornet Clearwing feed on the wood, living inside the twigs and trunk respectively. In the case of these 3 moth species, Goat Willow is the most important food plant, although their larvae will feed on other willow species. However, larvae of the Purple Emperor butterfly will feed on nothing else.

Purple Emperor caterpillar

Goat Willow foliage

Goat Willow catkins

Goat Willow bark

Lunar Hornet Clearwing

Goat Willow tree in summer

Goat Willow tree in winter

In addition to the sizeable willows (pp. 130–8), 7 low-growing native *Salix* species also occur in the British Isles. All but one are associated with northern and upland areas.

Creeping Willow *Salix repens* (Salicaceae) 1.5m

Low-growing and creeping shrub. SHOOTS Sometimes downy; usually reddish brown. LEAVES Ovate, usually untoothed, to 4cm long; hairless above when mature, with silky hairs below. REPRODUCTIVE PARTS Catkins. STATUS AND DISTRIBUTION Locally common on moors, heaths and coastal dune slacks.

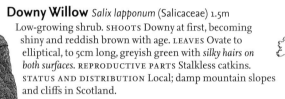

Downy Willow *Salix lapponum* (Salicaceae) 1.5m

Low-growing shrub. SHOOTS Downy at first, becoming shiny and reddish brown with age. LEAVES Ovate to elliptical, to 5cm long, greyish green with *silky hairs on both surfaces*. REPRODUCTIVE PARTS Stalkless catkins. STATUS AND DISTRIBUTION Local; damp mountain slopes and cliffs in Scotland.

Creeping
Willow
leaf

Downy
Willow
leaf

Woolly Willow *Salix lanata* (Salicaceae) 3m

Small shrub. SHOOTS Woolly at first, glossy brown with age. LEAVES Broadly oval, to 6cm long, white and woolly below when mature. REPRODUCTIVE PARTS Catkins; males golden. STATUS AND DISTRIBUTION Rare, on damp, base-rich mountain ledges.

Woolly
Willow
leaf

Mountain
Willow
leaf

Mountain Willow *Salix arbuscula* (Salicaceae) 1.5m

Small shrub. SHOOTS Dark brown and shiny when mature. LEAVES Ovate, to 4cm long, shiny green above but downy grey below. REPRODUCTIVE PARTS Catkins; males with red anthers. STATUS AND DISTRIBUTION Local, on base-rich mountain flushes.

Net-leaved Willow *Salix reticulata* (Salicaceae) 10cm

Mat-forming undershrub with distinctive leaves. SHOOTS Smooth and reddish brown when mature. LEAVES Ovate, to 5cm long and untoothed; dark green above but whitish below with prominent, netted veins. REPRODUCTIVE PARTS Catkins; males with reddish stamens. STATUS AND DISTRIBUTION Rare in Scottish Highlands.

Net-leaved
Willow
leaf

Dwarf
Willow
leaf

Dwarf Willow *Salix herbacea* (Salicaceae) 10cm

Prostrate, spreading undershrub. SHOOTS Smooth and reddish brown when mature. LEAVES Round, to 2cm long and toothed; shiny green above, pale below, both sides with obvious veins. REPRODUCTIVE PARTS Catkins. STATUS AND DISTRIBUTION Locally common on mountains and, further north, at lower altitudes.

Whortle-leaved Willow *Salix myrsinites* (Salicaceae) 50cm

Spreading undershrub. SHOOTS Glossy reddish brown when mature. LEAVES Oval, to 5cm long, shiny green with obvious veins on both sides. REPRODUCTIVE PARTS Catkins. STATUS AND DISTRIBUTION Rare, on base-rich mountain soils.

Whortle-leaved
Willow leaf
upperside

Whortle-leaved
Willow leaf
underside

Creeping Willow catkins

Creeping Willow plant and fruits

Downy Willow foliage

Woolly Willow foliage

Mountain Willow flowers and foliage

Net-leaved Willow flowers and foliage

Dwarf Willow foliage

Dwarf Willow plant

Whortle-leaved Willow foliage

POPLARS *POPULUS* (FAMILY SALICACEAE)

About 30 species, most of which grow rapidly to form large trees. Numerous hybrids occur.

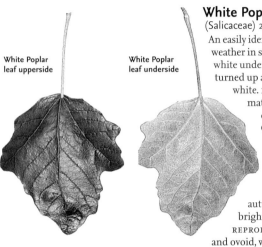

White Poplar
leaf upperside

White Poplar
leaf underside

White Poplar (Abele) *Populus alba*

(Salicaceae) 20m

An easily identified tree in windy weather in summer, when the pure white undersides of the leaves are turned up and the whole tree looks white. BARK White on the trunk and, in mature specimens, broken by diamond-shaped scars. SHOOTS Covered in white felt, which usually wears off by the end of the growing season. LEAVES Simple, deeply lobed and covered with dense white felt underneath; greyish green above. Leaves fall early in the autumn, sometimes turning a pleasing bright yellow for a few days before falling. REPRODUCTIVE PARTS Male catkins long and ovoid, white and fluffy; female catkins more slender and greenish yellow. STATUS AND DISTRIBUTION A native of mainland Europe, but presumed to have been an early introduction to Britain, perhaps because of its rapid rate of growth and its ability to flourish even in the poorest of soils and on the most exposed sites. Common near the coast, where it grows in thickets as a result of suckering, but less frequent inland. COMMENTS In the past, the timber was used for planking, and to make packing crates and toys. Trees growing in waterlogged settings are often host to the larvae of the Goat Moth, which burrow in, and eat, the wood.

Grey Poplar *Populus × canescens* (Salicaceae) 37m

Stable hybrid between White Poplar and Aspen. When fully mature, grows into an impressively large tree with a good solid bole. In spring the tree has a whitish appearance when the wind displaces the leaves, but it is not as brilliantly white as the White Poplar. BARK Whitish with diamond-shaped fissures. LEAVES Rounded to oval and toothed with regular, blunt, forward-pointing teeth; borne on long petioles. Leaf upper surface glossy grey-green; lower surface covered with greyish-white felt. By mid- to late summer the leaves lose some of the white felt and the tree looks greyer. REPRODUCTIVE PARTS Male and female catkins on separate trees. Female trees, with green, pendulous catkins are rare. Male catkins elongated and pendulous, giving the whole tree a purplish colour when they swell before opening in spring. STATUS AND DISTRIBUTION Native of mainland Europe and introduced to Britain very early, probably with the White Poplar. Our best specimens are found in chalky areas of S England and in limestone valleys of Ireland. However, it tolerates a wide range of climates and soil types and is found in N Scotland as well as in many coastal places. COMMENTS This tree suckers freely, so sometimes appears to be growing in thickets of the same species.

Grey Poplar
leaf

White Poplar foliage

White Poplar foliage

White Poplar bark

Goat Moth

Goat Moth larva

White Poplar

Grey Poplar foliage

Grey Poplar tree in winter

Grey Poplar bark

Aspen *Populus tremula* (Salicaceae) 18m

Slender to slightly conical tree with a rounded crown and tall, tapering trunk. Best known for its fluttering leaves, which rustle in the slightest breeze. BARK Smooth and greyish green at first, becoming brown, ridged and fissured with age. LEAVES Rounded to slightly oval, with shallow marginal teeth. Green on both surfaces, but paler below, on long, flattened petioles. In autumn, leaves may turn golden yellow, especially in the N; further S they often fall quickly without a colour change. Leaves newly produced in summer are often deep red. REPRODUCTIVE PARTS Catkins, to 8cm long, in clusters at ends of twigs, male and female on different trees. Male catkins reddish purple, females green tinged pink. Seeds are produced prolifically. STATUS AND DISTRIBUTION Native to Britain and common in many places, especially on poor, damp soils. COMMENTS Can easily be recognised by sound alone! The characteristic dry rustling of the leaves, especially on almost still days when no other leaves are moving, is quite distinctive. Aspen suckers readily, often growing in small groves of the same sex.

Aspen leaf

Black-poplar *Populus nigra* ssp. *betulifolia* (Salicaceae) 32m

Large, spreading tree when fully mature, with a domed crown and thick, blackish, gnarled bole covered with distinctive burrs and tuberous growths. Rather straggly, however, when growing among other trees in woods. BARK Grey-brown, darkening with age, becoming deeply fissured. SHOOTS (AND BUDS) Smooth and golden brown when young. LEAVES Triangular to diamond-shaped and variably long-stalked with a finely toothed margin; fresh shiny green on both surfaces. REPRODUCTIVE PARTS Male catkins pendulous and reddish, female catkins greenish. Both appear in April. STATUS AND DISTRIBUTION Native of the British Isles, preferring heavier soils and damp conditions. Tolerates pollution so sometimes planted in cities. COMMENTS Known in N England as 'Manchester Poplar'. The main host for the larvae of the Hornet Moth *Sesia apiformis*.

Black-poplar leaves

Lombardy-poplar *Populus nigra* 'Italica' (Salicaceae) 36m

Distinctive, narrowly columnar tree. Gnarled bole supports numerous short, ascending branches that taper towards narrow, pointed crown. Otherwise similar to Black-poplar, with slightly more triangular leaves. REPRODUCTIVE PARTS Typical, slender Lombardy-poplars are all males, bearing reddish catkins. Female trees of the clone 'Gigantea' are scarce and have thicker, spreading branches that form a broader crown. STATUS AND DISTRIBUTION Native of Italy, introduced to Britain in mid-18th century. Tolerates a wide range of soils and climates. Often planted in long lines.

SIMILAR TREE

Berlin Poplar *P.* × *berolinensis* (27m) More spreading and tolerant of pollution; planted occasionally.

Lombardy-poplar leaf

Aspen trunk

Aspen foliage

Aspen tree in autumn

Black-poplar bark

Black-poplar tree in winter

Black-poplar

Hornet Moth

Lombardy-poplar foliage

Lombardy-poplar

Hybrid Black-poplar *Populus* × *canadensis* (Salicaceae) 30m

Upright or spreading tree (depending on situation), with a narrow crown. Similar to Black-poplar, one of its parent species, and in many areas far more common; the other parent is the North American tree Cotton-wood. Trunk lacks the burrs seen in Black-poplar. BARK Deeply fissured and greyish. SHOOTS Young twigs greenish or slightly reddened. LEAVES Alternate, oval to triangular, and sharply toothed with fringes of small hairs. REPRODUCTIVE PARTS Catkins, similar to those of Black-poplar. STATUS AND DISTRIBUTION Planted for ornament or timber (used for packing crates and boxes). Does not thrive in wet or cold areas. COMMENTS Many Hybrid Black-poplar forms occur, separated by leaf structure and tree shape, e.g. Black Italian-poplar.

Hybrid Black-poplar leaf

Western Balsam-poplar *Populus trichocarpa* (Salicaceae) 35m

Fast-growing (up to 2m per year), columnar when mature with a tapering crown and trunk. BARK Dark grey with shallow grooves and fissures. SHOOTS Stout. LEAVES Pointed, tapering, glossy green above and white below, turning yellow in autumn. REPRODUCTIVE PARTS Catkins, produced in April, slender and pendulous: males reddish brown and females greenish. Seeds hairy and abundant. STATUS AND DISTRIBUTION Native of Pacific coast of North America, sometimes planted in Britain. COMMENTS Often planted in sites unsuited to its rapid growth and large size when mature. Small trees regenerate well when pruned.

Western Balsam-poplar leaf

Hybrid *Populus* 'Balsam-spire' leaf

SIMILAR TREE

Populus 'Balsam Spire' (35m) A hybrid between Western and Eastern Balsam-poplars, growing to a domed spire with a spreading crown. Bark silvery black and leaves triangular but rounded at the base. Widely planted.

Eastern Balsam-poplar *Populus balsamifera* (Salicaceae) 30m

Conical to slightly spreading tree with numerous ascending branches arising from a tapering bole with its base often surrounded by suckers. BARK Thinner than in other poplars and narrowly grooved. SHOOTS Young shoots (and 2.5cm-long buds) covered with shiny resin. LEAVES To 10cm long, oval and pointed at tip with finely toothed margins; dark shiny green above, paler and downy below. REPRODUCTIVE PARTS Greenish catkins appear in late spring or early summer. Males (to 7.5cm long) and females (to 12.5cm long) on separate trees. STATUS AND DISTRIBUTION Native of North America. Cultivated elsewhere occasionally. Cultivated trees appear to be all males.

Eastern Balsam-poplar leaf

SIMILAR TREE

Balm-of-Gilead *P.* × *jackii* (20m) Buds abundant and sticky, on downy shoots that are balsam-scented. Young leaves aromatic when newly opened, heart-shaped and downy below; on a downy petiole. Mature tree open-crowned, suckering freely and forming thickets.

Balm-of-Gilead leaf

Hybrid Black-poplar foliage

Hybrid Black-poplar bark

Western Balsam-poplar bark

Hybrid Black-poplar

Western Balsam-poplar

Western Balsam-poplar foliage

Eastern Balsam-poplar foliage

Balm-of-Gilead bark

Balm-of-Gilead foliage

HICKORIES *CARYA* (FAMILY JUGLANDACEAE)

19 species occur in North America, 1 in SW China. Male catkins in 3 parts. Internal pith is solid.

Shagbark
Hickory
leaf

Shagbark Hickory *Carya ovata* (Juglandaceae) 20m

Upright or slightly spreading tree with a broad, flattened crown. BARK Grey, splits into long scaly flakes. BRANCHES In winter, sparse branches support reddish twigs tipped with scaly buds. LEAVES Compound, 5 (sometimes 3 or 7) leaflets, to 20cm long, longer near tip of leaf. Leaflets oval to oblong, toothed, with tufts of short white hairs between teeth. Terminal leaflet short-stalked. Leaves leathery, sometimes oily. REPRODUCTIVE PARTS Male catkins green, to 15cm long, in spreading clusters. Female flowers small, yellowish, in terminal clusters. Fruits round, to 6cm long, on short stalks; containing white seeds. STATUS AND DISTRIBUTION Native of E North America, planted in Britain occasionally.

Shagbark
Hickory
fruit

Bitternut *Carya cordiformis* (Juglandaceae) 30m

Bitternut
leaf

Bitternut
fruit

Large tree with a high conical crown. BARK Greyish, smooth at first becoming scaly with age; peeling flakes revealing orange patches beneath. BRANCHES Mostly straight and ascending, the greenish twigs tipped with elongated, yellowish and scaly buds. LEAVES Compound, with 9 leaflets (rarely 5 or 7); terminal leaflet stalkless. Individual leaflets elongated, pointed at tip with toothed margins. REPRODUCTIVE PARTS Male catkins, to 7cm long, yellowish and pendulous. Fruits, to 3.5cm long, rounded to pear-shaped with 4 wings, concealing grey, smooth seeds. STATUS AND DISTRIBUTION Native of E North America; planted in Britain mainly for ornament.

Caucasian
Wingnut
fruits

SIMILAR TREE

Pignut *C. glabra* (20m) A fine specimen tree, but rarely planted. Bark smooth; leaves like Ash.

WINGNUTS *PTEROCARYA* (FAMILY JUGLANDACEAE)

Attractive, suckering trees. Winter buds protected by 2 closely adpressed hairy leaves. Twig pith is divided into chambers. Winged seeds in pendulous catkins.

Caucasian
Wingnut
leaf

Caucasian Wingnut *Pterocarya fraxinifolia* (Juglandaceae) 35m

Spreading tree with domed crown and stout bole from which many branches arise close to the same point. Suckers freely. BARK Grey, fissured and gnarled. LEAVES Compound, with up to 20 pairs of leaflets, each to 18cm long, ovate to lanceolate with a pointed tip and toothed margins. Midribs bear stellate hairs on underside. Leaves turn yellow in autumn. REPRODUCTIVE PARTS Male catkins solitary, female catkins pendent with many flowers, giving rise to broad-winged nutlets. STATUS AND DISTRIBUTION Native of SW Asia, planted elsewhere for ornament.

SIMILAR TREE

Chinese Wingnut *P. stenoptera* (25m) Leaves compound with toothed wings on central blade; fewer leaflets. Winged fruits pink.

Caucasian Wingnut fruit

Shagbark Hickory fruits and foliage

Shagbark Hickory bark

Caucasian Wingnut foliage

Bitternut foliage

Caucasian Wingnut fruits and foliage

Caucasian Wingnut bark

WALNUTS *JUGLANS* (FAMILY JUGLANDACEAE)

15 species exist, from North and South America and Asia. They produce large edible nuts and excellent timber. Twig pith is chambered. Can be propagated from seed and grafting.

Common
Walnut leaf

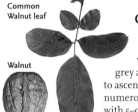

Walnut

Common Walnut *Juglans regia* (Juglandaceae) 30m

Spreading, deciduous tree. Has domed crown and straight bole when grown in ideal conditions; often contorted when found in orchards. BARK Smooth, brown at first, grey and fissured with age. BRANCHES Lowest ones spreading to ascending, often large near base, but dividing rapidly into numerous twisted twigs with dark purple-brown buds. LEAVES Compound, with 5–9 elliptical leaflets, to 15cm long, thick and leathery, with pointed tips and untoothed margins. Crushed leaves slightly aromatic. Opening late in spring, reddish, becoming green later in summer. REPRODUCTIVE PARTS Male catkins yellow, to 15cm long; female flowers small, greenish with yellow, protruding, branched stigma. Fruits rounded, to 5cm long; smooth green skin, dotted with slightly raised glands, encasing familiar edible walnut seed. STATUS AND DISTRIBUTION Native of SE Europe and Asia, long cultivated and may have arrived in Britain with the Romans. Rarely naturalised.

Black
Walnut
leaf

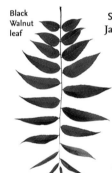

SIMILAR TREE

Japanese Walnut *J. ailanthifolia* (20m) Leaves much larger with 11–15 pointed, toothed leaflets, hairy on both surfaces, on hairy shoots. Fruits in clusters of up to 20.

Black Walnut *Juglans nigra* (Juglandaceae) 32m

Has a tall, straight bole and domed crown of brighter green leaves than Common Walnut. BARK Dark brown, showing a diamond pattern of deep cracks. LEAVES Compound with 15–23 leaflets, finely toothed and downy below. REPRODUCTIVE PARTS Fruits similar to Common Walnut but not as edible; green husk yields a similar dark dye. STATUS AND DISTRIBUTION Native of USA, planted in Britain occasionally.

Butternut
leaf

Butternut *Juglans cinerea* (Juglandaceae) 26m

Slender tree. BARK Grey. LEAVES Compound, to 70cm long, leaflets more widely spaced than in Black Walnut; central leaf stalk densely hairy and leaflets near leaf base smallest. REPRODUCTIVE PARTS Edible fruits in clusters of up to 12. STATUS AND DISTRIBUTION Native of E North America, planted elsewhere for ornament.

Butternut
bark

BOG MYRTLES *MYRICA* (FAMILY MYRICACEAE)

Shrubs or small trees, sometimes with aromatic leaves, and flowers in catkins.

Bog
Myrtle
leaf

Bog Myrtle *Myrica gale* (Myricaceae) 1m

Woody, brown-stemmed deciduous shrub. LEAVES Oval, grey-green and resin-scented. REPRODUCTIVE PARTS Orange, ovoid male catkins and pendulous brown female catkins, on separate plants. Fruits are brownish nuts. STATUS AND DISTRIBUTION Widespread but local on boggy heaths.

Common Walnut fruit

Common Walnut

Black Walnut fruit and foliage

Black Walnut fruit and foliage

Black Walnut

Butternut flowers

Butternut flowers and foliage

Bog Myrtle flowers and foliage

BIRCHES *BETULA* (FAMILY BETULACEAE)

About 40 species occur across the whole of the northern hemisphere, some surviving further N than any other tree species. Most grow into medium-sized trees with good hard timber used for making plywood, and sometimes for paper pulp. Few birches grow large enough to provide sizeable beams or planks for building. Flowers are in the form of catkins and the seeds are very small winged nutlets dispersed by the wind over great distances and in large numbers. They are rapid colonisers of disturbed ground, and some birches are very invasive. The seeds are important as a winter food for small birds, such as Siskins and Redpolls, and the leaves are food for innumerable insect larvae.

scale

leaf

seed

Silver Birch *Betula pendula* (Betulaceae) 26m

A slender, fast-growing deciduous tree with a narrow, tapering crown when young and growing vigorously. Older trees acquire a weeping habit, especially if growing in an open, uncrowded situation. BARK In old trees, thick, deeply fissured at the base of the bole, breaking up into rectangular plates; higher up the bole the bark is a smooth silvery white, often flaking away and revealing greyer patches below. A pattern of black diamond shapes is often seen on the trunk of older trees. BRANCHES Ascending in young trees, but twigs and shoots pendulous, slender and smooth, mostly brown and pitted with many white resin glands. LEAVES Up to 7cm long, triangular and pointed with large teeth separated by many smaller teeth. Thin and smooth when mature, and borne on hairless petioles. Turning golden yellow in autumn. REPRODUCTIVE PARTS Male catkins in groups of 2–4 at the tips of young twigs, appearing very early in the winter, when they are brownish in colour; yellow and pendulous in spring, when the leaves are opening. Female catkins shorter, more erect, greenish, produced in the axils of leaves. After pollination they become browner and thicker, eventually reaching a length of 3.5cm. Seeds winged and papery and usually produced copiously. STATUS AND DISTRIBUTION A native of a wide area of Europe, including Britain and Ireland. Often planted as an ornamental tree in gardens and parks. In the wild, often colonises heathland areas and its invasive habit is usually not welcomed, trees typically being felled and cleared. COMMENTS A fast-growing tree and an early coloniser, although it does not thrive in shade or compete with even more vigorous species. Silver Birch timber is popular as firewood. As a habitat, Silver Birch woodland is rich in fungi, some of which are associated with the tree almost exclusively. These include Fly Agaric *Amanita muscaria*, Brown Birch Bolete *Leccinum scabrum*, Woolly Milk Cap *Lactarius torminosus*. The Birch Polypore *Piptoporus betulinus*, a bracket fungus, grows on trunks.

LEFT: **Brown Birch Bolete**

BELOW: **Birch Polypore**

Silver Birch catkins

Silver Birch trees in autumn

Silver Birch

Silver Birch bark

Downy Birch *Betula pubescens* (Betulaceae) 25m

An attractive and rather elegantly proportioned tree when growing out in the open. It is superficially similar to Silver Birch, often growing alongside it. The crown in winter looks untidy compared with Silver Birch. It is a rather variable species but usually easy to recognise by the soft, downy feel of the tips of the twigs in spring, and by the reddish bark on young wood. BARK Mostly smooth and brown or greyish, but not breaking up into rectangular plates at the base like that of Silver Birch. BRANCHES More irregular and densely crowded than Silver Birch, and mostly erect, never pendulous. The twigs lack the whitish resin glands, but do have a covering of downy white hairs. LEAVES More rounded at the base than Silver Birch and more evenly toothed; white hairs in the axils of the veins on the underside; petiole hairy. REPRODUCTIVE PARTS Catkins very similar to Silver Birch but the winged seeds have smaller wings, about the same size as the seed itself. Female flowers in April–May. STATUS AND DISTRIBUTION A native of most of Europe, including Britain and Ireland, and common in some areas where soils are poor or peaty; common and widespread in upland areas, and in the west and north. COMMENTS Downy Birch is a rapid coloniser of cleared and recently burned areas and is often invasive on heathland. The felled timber is good for firewood and, like Silver Birch, is often used in wood-turning. The leaves are the food plant for large numbers of insect species, many moths among them. As with Silver Birch, a wide range of soil fungi are associated with Downy Birch woodland. In the Scottish Highlands, the Hoof Fungus *Fomes fomentarius*, a tough bracket fungus, is often found growing on stumps and trunks.

Downy Birch leaf

Downy Birch bark, young tree

Downy Birch bark, old tree

Dwarf Birch *Betula nana* (Betulaceae) 1m

A low-growing and often rather prostrate undershrub. BARK Reddish brown. BRANCHES Short and upright or spreading, with stiff, hairy twigs. LEAVES Rather rounded, 6–8mm across, coarsely toothed and hairy when young, smooth and hairless when mature. REPRODUCTIVE PARTS Catkins, females covered with 3-lobed scales. STATUS AND DISTRIBUTION A northern, tundra species in global terms, confined in Britain as a native plant to the Scottish Highlands, where it grows on upland heaths and in bogs and is tolerant of both waterlogged and relatively free-draining peaty soils.

Dwarf Birch leaf

actual size

Downy Birch foliage

Downy Birch catkins and foliage

Downy Birch

Hoof Fungus

Dwarf Birch catkins and spring foliage

Dwarf Birch autumn foliage

Paper-bark Birch bark

Paper-bark Birch (Canoe-bark Birch)
Betula papyrifera (Betulaceae) 23m
A stout, spreading tree. Best recognised by its leaves. BARK Mostly white and smooth, flecked with grey or sometimes orange or brown; peeling horizontally into strips. BRANCHES Spreading, the shoots covered in rough warts and a few long hairs. LEAVES Large by birch standards (to 10cm long), dull green and with only 5 pairs of veins; borne on hairy stalks. REPRODUCTIVE PARTS Catkins, females eventually producing winged seeds. STATUS AND DISTRIBUTION A native of N North America from the E to the W coasts. It is planted in Britain as an ornamental tree, mainly for the novelty of its bark. COMMENTS The freely peeling bark was once used by Native Americans to make canoes.

Paper-bark Birch leaf

Himalayan Birch bark

Himalayan Birch *Betula utilis* (Betulaceae) 20m
An elegant tree, rather rounded in specimens found growing in the open. BARK Extremely colourful, gleaming white in var. *jacquemontii* but pink, red or golden in other forms. The bark is marked with horizontal lenticels and peels off horizontally into rolls. BRANCHES Mainly upright rather than spreading, with twigs that are hairy when young. LEAVES Oval, with a pointed tip and toothed margins; dark green with 7–14 pairs of veins depending on the variety. REPRODUCTIVE PARTS Catkins: males long, pendulous and yellow. STATUS AND DISTRIBUTION A native of the Himalayas. Planted in Britain as an ornamental tree, mainly for its bark, which is particularly striking and evident on a sunny winter's day.

Himalayan Birch leaf

Erman's Birch *Betula ermanii* (Betulaceae) 24m
A fast-growing and attractive tree, easily recognised by its bark. The tree is more spreading and has a stouter bole than Silver Birch, with which it frequently hybridises. BARK Pinkish, or sometimes shining yellowish white; peeling horizontally and hanging in tattered strips down the bole of mature trees; younger trees have smoother, white bark. BRANCHES Rather upright, with twigs that are warty and usually hairless. LEAVES Triangular to heart-shaped with a pointed tip, toothed margins and 7–11 pairs of veins; stalks hairless. REPRODUCTIVE PARTS Catkins. STATUS AND DISTRIBUTION A native of E Asia, introduced into Britain and Ireland from Japan. The best specimens are seen in established gardens where the peeling bark is a fine winter feature.

Erman's Birch leaf

Himalayan Birch foliage

Himalayan Birch bark peeling

Paper-bark Birch tree in autumn

Erman's Birch foliage

Erman's Birch bark

ALDERS *ALNUS* (FAMILY BETULACEAE)

A genus of about 30 species, mostly found in wet habitats and especially characteristic of river banks. The roots have numerous bacteria-containing nodules to fix nitrogen, like members of the Pea family, so they can cope well on infertile soils. The clusters of bright orange nodules are often exposed when the roots of riverside trees are seen at low water levels. Alders are useful in protecting river banks from erosion and provide valuable cover for riverside wildlife. Seeds are borne in small, woody, cone-like catkins. The durable timber can withstand alternate wetting and drying.

Common Alder, floating seeds

Common
Alder cones

Common
Alder leaf

Common Alder *Alnus glutinosa*

(Betulaceae) 25m

A small, spreading and sometimes multi-stemmed tree with a broad domed or conical crown. BARK Brownish and fissured into square or oblong plates. BRANCHES Ascending in young trees, but spreading later. Twigs smooth except when young, when they have a sticky feel (hence *glutinosa*), with raised orange lenticels. Buds about 7mm long, on stalks 3mm long. LEAVES Stalked and noticeably rounded, up to 10cm long, with a slightly notched apex and a wavy or bluntly toothed margin. The 5–8 pairs of veins have long hairs in the axils on the underside of the leaf. REPRODUCTIVE PARTS In winter the purplish male catkins, in bunches of 2–3, are an attractive feature, even though they are only around 3cm long; by the end of winter they open up, revealing yellow anthers, and are more colourful. Female catkins are smaller (1.5cm) and cone-like, reddish purple at first and then turning green, usually in bunches of 3–8. They form hard green 'cones', which grow through the summer and persist until the following spring. Their small winged seeds float on water, which aids their dispersal. They are also an important source of winter food for finches like Redpolls and Siskins. STATUS AND DISTRIBUTION A native species throughout Europe, including Britain and Ireland, absent only from the very far north. It commonly grows beside water and can be found at altitudes up to 700m. COMMENTS Wetland-colonising alder woodlands are referred to as *carr*. In locations where it is left to its own devices, Common Alder forms an elegant, domed tree. However, in many places it is, or has been, coppiced to produce long, straight poles, creating broad, spreading, multi-stemmed boles. Common Alder is a useful timber for wet situations, so it is used for pier pilings, lock gates and making clogs. In the past, it was reckoned to be the best source of charcoal for gunpowder. The wood has an attractive bright orange colour when freshly cut. Growing trees help stabilise river banks and prevent erosion with their tough roots. The leaves are the food plant for many insects, particularly the larvae of moths. Larvae of the White-barred Clearwing *Synanthedon spheciformis* feed inside the trunk.

Common Alder bark

Common Alder cones and foliage

Common Alder carr

White-barred Clearwing

Common Alder, freshly cut timber

Green Alder
leaf

Green Alder *Alnus viridis* (Betulaceae) 5m

Rarely more than a large shrub or small tree. BARK Brown. BRANCHES With mostly smooth, greenish twigs and pointed, sessile, shiny red buds. LEAVES More pointed than those of Common Alder and sharply toothed, hairy on the midrib and in the joins of the veins on the underside. When first open they are sticky to the touch. REPRODUCTIVE PARTS Male catkins, appearing with the leaves, are up to 12cm long, yellow and pendulous; female catkins are 1cm long, erect and greenish at first, becoming reddish later, and usually found in stalked clusters of 3–5. The cone-like ripe catkins are rounded, green and tough at first, becoming blackened later and persisting until the following spring. STATUS AND DISTRIBUTION Native of the mountains of central and E Europe, planted in Britain for ornament.

Grey Alder *Alnus incana* (Betulaceae) 25m

A fast-growing alder more at home on dry soils than most other alders. Shoots and new leaves are covered with a dense layer of soft greyish hairs (hence *incana*). BARK Smooth and grey. LEAVES Triangular and toothed, terminating in a point; margins not rolled inwards. Hairs persisting on underside of leaf as it matures. REPRODUCTIVE PARTS Catkins and fruits very similar to other alders, although the green fruits are more globose before ripening to the typical dark, woody alder cone. STATUS AND DISTRIBUTION A native of Europe, introduced to Britain but not often planted. A good species for wasteland and reclamation schemes.

Grey Alder
leaf

SIMILAR TREES

Red Alder *A. rubra* (25m) Resembling Grey Alder, but toothed margin of leaf inrolled (check with a hand-lens). Leaves up to 20cm long. Catkins and cones very similar to those of Common Alder. A native of W North America. Not often planted in Britain. However, where it does occur, such as in Scotland, it grows rapidly at first, then slows down.
Smooth Alder *A. rugosa* (20m) A North American species with red hairs in axils of leaf veins on the singly toothed leaves. Planted occasionally.

Italian Alder
leaf

Italian Alder *Alnus cordata* (Betulaceae) 29m

An attractive tree with a bold, conical shape, fine glossy leaves and an impressive show of catkins and cones. BARK Pale grey and fairly smooth with slightly downy twigs. LEAVES The best feature for identification: glossy, heart-shaped (hence *cordata*), with short tufts of orange hairs along the midrib on the underside. REPRODUCTIVE PARTS Male catkins yellow and produced prolifically; female catkins in small clusters, ripening in early summer. Woody 'cones' larger than those of any other alder species. STATUS AND DISTRIBUTION Native of Corsica and S Italy, planted in British parks and gardens, and often along roadsides.

Green Alder
catkins and foliage

Grey Alder foliage

Grey Alder bark

Grey Alder cones

Italian Alder catkins, cones and foliage

Italian Alder
catkins

Italian Alder cone

HORNBEAMS *CARPINUS* **AND HOP-HORNBEAMS** *OSTRYA* (FAMILY BETULACEAE)

About 45 species occur in the northern hemisphere; related to birches and alders but sometimes placed in their own family, Carpinaceae. Male catkins are protected inside winter buds. Leaves are sharply toothed and have conspicuous parallel veins.

ABOVE: **Hornbeam bark**
RIGHT: **Hornbeam leaf**

RIGHT:
Hornbeam fruit
BELOW: **Oriental Hornbeam fruit**

Hornbeam *Carpinus betulus* (Betulaceae) 30m

A fine tree with a bold outline in winter. Bole is often gnarled and twisted. BARK Silvery grey with deep fissures lower down and occasional dark bands. BRANCHES Densely packed, ascending and twisted, bearing greyish-brown, partly hairy twigs. LEAVES Oval and pointed with a rounded base, short petiole, and double-toothed margin; 15 pairs of veins hairy on underside. Colourful in autumn, turning yellow, through orange to russet-brown; trees planted in hedgerows retain leaves long into the winter. REPRODUCTIVE PARTS Male catkins, to 5cm long, yellowish green with red outer scales. Fruits in clusters of winged nutlets, to 14cm long, usually consisting of about 8 pairs of small hard-cased nuts with a 3-pointed papery wing. STATUS AND DISTRIBUTION Native to Britain, occurring in pure stands in some woodlands and hedgerows. Also widely planted and seen as a specimen tree in parks and gardens. Tolerant of heavy clay soils. COMMENTS Hornbeam was regularly coppiced in the past to provide a timber crop. The tough wood is prized for its durable qualities: it was used for wheel hubs, mill-wheels, piano hammers and chopping blocks. The tough seeds are a favourite food of the Hawfinch, the only British bird able to crack them open.

Oriental Hornbeam *Carpinus orientalis* (Betulaceae) 11m

Similar to Hornbeam but separable with care; overall the tree is normally smaller and neater. BARK Like Hornbeam. BRANCHES With thinner shoots than Hornbeam, covered with long silky hairs. LEAVES Like Hornbeam but smaller, always looking slightly folded. REPRODUCTIVE PARTS Similar to Hornbeam but fruit bracts unlobed (3-lobed in Hornbeam). STATUS AND DISTRIBUTION Native of SE Europe and Asia Minor, occasionally seen in Britain as a specimen tree.

RIGHT:
Oriental Hornbeam leaf

European Hop-hornbeam *Ostrya carpinifolia* (Betulaceae) 19m

A spreading tree with a domed crown and robust bole. BARK Grey-brown, with squarish plates. BRANCHES Almost level when growing in the open, but in woodland may be crowded and ascending. LEAVES Like Hornbeam. REPRODUCTIVE PARTS Fruits in clusters with a superficial resemblance to bunches of hops. STATUS AND DISTRIBUTION Native of mainland Europe; in Britain it occurs mainly in well-established gardens.

European Hop-hornbeam fruit

European Hop-hornbeam leaf

Hornbeam fruits and foliage

Hornbeam pollarded trees

Hornbeam catkins

Hornbeam fruits

European Hop-hornbeam catkins and young leaves

Oriental Hornbeam fruits and foliage

European Hop-hornbeam fruits and foliage

HAZELS *CORYLUS* (FAMILY BETULACEAE)

About 15 species, of which only 4 reach tree status, all confined to northern hemisphere. Prominent male catkins open early in winter, and female flowers are little more than tiny buds. Fruits are edible hard-shelled nuts.

Hazel leaf

Hazel *Corylus avellana* (Betulaceae) 6m

Often no more than a spreading, multi-stemmed shrub, but sometimes grows into a taller tree with a shrubby crown and a short but thick and gnarled bole. BARK Smooth and often shiny, peeling horizontally into thin papery strips. BRANCHES Upright to spreading, depending partly on management regime. Twigs covered with stiff hairs; buds oval and smooth. LEAVES Rounded, to 10cm long, with a heart-shaped base and pointed tip.

Hazel nut

Margins double-toothed and upper surface hairy. On underside, white hairs on leaf veins. Petiole short and hairy; whole leaf has a bristly, rough feel. REPRODUCTIVE PARTS Male catkins first appear in autumn and are short and green, but when they open early in spring they are up to 8cm long, pendulous and yellow. Female flowers red and very small, producing hard-shelled nuts in bunches of 1–4; nuts partly concealed in a leafy, deeply toothed involucre. Nuts up to 2cm long, brown and woody when ripe. STATUS AND DISTRIBUTION A widespread native tree across most of Europe, including Britain and Ireland, occurring in hedgerows and woodlands where it is an important

Hazel nuts

component of the understorey. COMMENTS Frequently coppiced to provide poles for a variety of uses and of immense importance to woodland wildlife for its edible leaves and fruits. The Dormouse *Muscardinus*

Filbert leaf

avellanarius is particularly reliant on the nuts and numerous moth larvae eat the leaves. Some Hazels that have been coppiced and recoppiced many times are now extremely old trees, having greatly exceeded their normal lifespan through the constant regeneration caused by cutting them back and allowing them to regrow.

Filbert nut

Filbert *Corylus maxima* (Betulaceae) 6m

Very similar to Hazel except for the nuts, which are longer, mostly solitary or in bunches of 2–3 and entirely enclosed in an undivided involucre, which is constricted over the nut and toothed at the tip. A native of the Balkans, but widely planted elsewhere for the superior quality of its nuts, and sometimes naturalised. 'Purpurea' is a commonly planted cultivar.

Filbert nuts

Turkish Hazel *Corylus colurna* (Betulaceae) 22m

Larger than Common Hazel with a stout bole and a conical crown. The best feature for identification is the involucre, which completely encloses the nut and is finely toothed and often recurved. Leaves are similar to those of Common Hazel but are more likely to look lobed. A native of SE Europe and Asia Minor, and also found as an introduction further N and W.

Turkish Hazel fruit (left) and leaf (right)

Hazel bark

Hazel foliage

Hazel, mass catkins

Hazel var. *contorta*

Hazel male catkins

Hazel female catkins

Filbert foliage

Turkish Hazel bark

Turkish Hazel catkins

BEECHES *FAGUS* (FAMILY FAGACEAE)

Large, impressive trees with smooth bark. Male flowers appear in rounded clusters and the 1 or 2 nuts are in woody and sometimes spiny husks.

Beech *Fagus sylvatica* (Fagaceae) 40m

Beech leaf

A large and imposing deciduous tree with a broad, rounded crown. BARK Usually smooth and grey, but may occasionally become rougher. BRANCHES Often crowded and ascending, but sometimes arching outwards. Buds up to 2cm long, smooth and pointed, and reddish brown. LEAVES To 10cm long, oval and pointed, with a wavy margin and a fringe of silky hairs when freshly open. Petiole up to 1.5cm long. REPRODUCTIVE PARTS Male flowers pendent, in clusters at tips of twigs. Female flowers paired, on short stalks, and surrounded by a brownish, 4-lobed involucre. Nuts up to 1.8cm long, 3-sided, shiny and brown, and enclosed in a prickly case in pairs.

Beech nut

Beech mast

STATUS AND DISTRIBUTION A widespread native of W and central Europe. It was one of the last native trees to colonise Britain and its natural range lies S and E of a line between the Wash and the Severn; widely planted elsewhere as an ornamental tree, in shelter-belts and hedgerows, and for timber. Beech prefers drier soils such as chalk, but is found on a wide variety of free-draining soils. COMMENTS The timber is mostly used for making furniture; it does not last well out of doors. Beech trees cast such a dense shade and produce such copious leaf litter that comparatively little grows beneath them and no other tree species can compete. However, a select group of specialist plants find Beech woodlands to their liking, notably helleborines (orchids) of the genera *Epipactis* and *Cephalanthera*, which favour clearings, and the saprophytic, shade-tolerant Bird's-nest Orchid *Neottia nidis-avis*. Comparatively few insects feed on the leaves of Beech, a notable exception being the larva of the Lobster Moth *Stauropus fagi*. By contrast, a wealth of fungi appear in autumn and species such as the Death Cap *Amanita phalloides*, Satan's Bolete *Boletus satanus*, the Porcelain Fungus *Oudemansiella mucida* and the Artist's Bracket *Ganoderma applantum* are reasonably Beech-specific. During the winter months, Bramblings and Chaffinches feed on the fallen mast.

SIMILAR TREES

Copper Beech *F. sylvatica* 'Purpurea' (40m) A densely purple tree, often looking rather overbearing and casting a very dense shade. Leaves shaped like those of Common Beech, but new leaves red and older leaves a deep opaque purple.

Dawyck Beech *F. sylvatica* 'Dawyck' (40m) A columnar form of Beech discovered in Dawyck, Scotland, in the mid-19th century. Resembling Lombardy Poplar from a distance, but more densely branched with otherwise normal Beech leaves, flowers and fruits. Still rare, but sometimes seen in parks and gardens and on roadsides.

Bird's-nest Orchid

Beech bark

Beech, winter woodland

Beech fruit and foliage

Beech, autumn colours

Beech flowers and foliage

Death Cap

Copper Beech

Oriental Beech *Fagus orientalis* (Fagaceae) 23m

Oriental Beech leaf

Leaves larger than those of Common Beech and widely separated, with 7 or more pairs of veins. A native of the Balkans, Asia Minor and the Caucasus, rare elsewhere. Grows vigorously and forms a fine tree in good conditions.

SOUTHERN BEECHES NOTHOFAGUS (FAMILY FAGACEAE)

About 40 species occur in South America and Australasia. A few are deciduous, but most are evergreen with fine, glossy foliage.

Rauli *Nothofagus procera* (Fagaceae) 28m

Rauli leaf

Attractive, conical tree with a stout bole and striking autumn foliage. BARK Grey, with vertical plates. BRANCHES Lower branches usually level, upper branches more ascending. Thick, green twigs, darkening with age; buds about 1cm long, pointed and reddish brown. LEAVES Alternate, to 8cm long and rather pointed at tip. Margin wavy, minutely toothed; 15–22 pairs of veins covered on underside with fine silky hairs. REPRODUCTIVE PARTS Male and female flowers on same tree. Males solitary, in leaf axils; female flowers also usually solitary, giving rise to 4-lobed hairy capsules containing 3 shiny brown nuts. STATUS AND DISTRIBUTION Native of Chile, introduced to Britain early in 20th century and found in parks, gardens and commercial plantations. Grows rapidly at first, up to 2m a year, and soon makes an attractive specimen tree.

Roble *Nothofagus obliqua* (Fagaceae) 30m

Roble leaf

More delicate in appearance than Rauli. BARK Silvery grey with curling plates. BRANCHES Slender, ascending branches and pendent shoots on upper crown. Twigs finer than Rauli, branching in a regular, alternate pattern. LEAVES 7–11 pairs of veins (compared with Rauli's 15–22) and a wavy margin. REPRODUCTIVE PARTS Flowers grow in leaf axils and fruits are 4-lobed hairy capsules. STATUS AND DISTRIBUTION Native of Chile and W Argentina; grown in the British Isles for ornament, occasionally for timber. COMMENTS Grows very fast, comes into leaf later than Rauli, and has good autumn colour.

SIMILAR TREES

Coigue *N. dombeyi* (28m) Evergreen, but rather tender until well established. Bark of young trees smooth and black, but becoming wrinkled and browner with age, with scales peeling away to leave red patches.

Coigue leaf

Antarctic Beech leaf

Antarctic Beech *N. antarctica* (16m) First discovered in its native Chile and Tierra del Fuego in the 1830s, and grown in Britain since then. Hardy, but prefers some shelter. An attractive small tree with delicate, shiny foliage and reddish, shiny bark in young trees. Leaves with only 4 pairs of veins and remaining curled for most of the season, turning a pleasing yellow and then brown in autumn.

Oriental Beech foliage

Rauli fruits and foliage

Roble flowers and foliage

Coigue fruits and foliage

Antarctic Beech fruits and foliage

Oriental Beech bark

Rauli bark

Roble bark

Coigue bark

SWEET CHESTNUTS CASTANEA (FAMILY FAGACEAE)

10 species exist and are found in the northern hemisphere, most coming from southerly areas that experience a temperate climate. Most species eventually grow to become large trees and have long leaves and edible nuts.

Sweet Chestnut leaf

Sweet Chestnut *Castanea sativa* (Fagaceae) 35m

A handsome, large-leafed deciduous tree with a fine bole and attractive autumn colours. BARK Silvery and smooth in young trees with fine vertical fissures, but in older trees the bark becomes more deeply fissured and the grooves markedly spiralled up the trunk. BRANCHES In old trees the lowest branches are often very large and spreading, the upper ones more ascending and twisted. LEAVES Glossy, up to 25cm long, lanceolate, the margins serrated with spine-tipped teeth, pointed at the tip, and sometimes with a slightly heart-shaped base. REPRODUCTIVE PARTS Male catkins creamy white, long and pendulous, producing a sickly sweet smell. The female flowers, in groups of 2 and 3 at the base of the male catkins, are greenish and erect, and give rise to the familiar spine-covered green capsules that split open to reveal 3 shiny, brown-skinned nuts. Flowers are produced in June–July, and the edible fruits are ripe in October–November. STATUS AND DISTRIBUTION A widespread native of most of mainland Europe and N Africa, but not native to the British Isles, where it was introduced by the Romans and is now frequently planted, and sometimes naturalised too. It thrives on a range of soils, but does especially well on well-drained, slightly acidic ground, and on hillsides. COMMENTS The edible nuts were a staple part of the diet of Roman legionaries, which explains the species' introduction to Britain. However, Sweet Chestnut seldom produces a significant crop of nuts in the British Isles, and today the timber is of greater economic significance. The wood is strong and durable, and Sweet Chestnut is frequently planted and managed commercially to this end: typically trees are coppiced on a 10- to 12-year cycle, generating tall, straight poles. Short lengths are then split lengthways to make *pales* and these are used in fencing. Thin, steam-bent planks are traditionally used to make Sussex trugs. Sweet Chestnut is of comparatively little value to native wildlife and relatively few moth larvae, for example, will feed on its foliage.

Sweet Chestnut leaf in autumn

Sweet Chestnut fruit

Sweet Chestnut bark

Sweet Chestnut fruit and foliage

Sweet Chestnut trunk

Sweet Chestnut fallen fruit

Sweet Chestnut foliage

OAKS *QUERCUS* (FAMILY FAGACEAE)

About 500 species exist in the northern hemisphere, many of them occurring in warmer climates (for example, Mexico has about 125 species). Some species hybridise freely, and a number of long-established, named hybrids exist. Around half are evergreens. Some of the deciduous oaks produce brilliant autumn colours. All reproduce by means of acorns. Many are fine and imposing trees producing high-quality, long-lasting timber and are of considerable commercial significance. Oaks are often slow-growing but long-lived, and some are immensely important to wildlife for food and shelter, dominating the landscape in many areas.

Pedunculate
Oak leaf

Pedunculate (English) Oak *Quercus robur* (Fagaceae) 36m

A large, spreading, deciduous tree with a dense crown of heavy branches. BARK Grey, becoming thick and deeply fissured in mature trees. BRANCHES Very old trees (700–800 years) may have dead branches emerging from upper canopy (giving rise to description 'stag-headed') and a hollow trunk. Shoots and buds hairless. LEAVES Deeply lobed with 2 auricles at the base; on very short stalks (5mm or less). First flush of leaves is often eaten rapidly by insects, and is replaced by a second crop in midsummer – so-called 'Lammas growth'. REPRODUCTIVE PARTS Male and female catkins produced just as first flush of leaves appears in spring. Male catkins die off after pollination, by which time leaves are fully open. Acorns are borne on long stalks in roughly scaled cups, in groups of 1–3. STATUS AND DISTRIBUTION Widespread native tree in Britain and Ireland, preferring heavier clay soils than the superficially similar Sessile Oak. Often dominant in old woodlands, especially in lowland areas, but it occurs in more hilly country as well. In almost all settings in the British Isles, mature oak woodland is classed as *semi-natural* because of the degree to which it has been, and still is, managed by man. COMMENTS Oak is an extremely important building material and for traditional timber-framed buildings it was certainly the timber of choice, and often the only one used. Little or no distinction was made between timber derived from our 2 native oak species, but in much of lowland Britain wood from Pedunculate Oak would have predominated, simply because it was more readily available. English Oak is a former name for this species; confusingly, today timber from both native oaks is sold as 'English oak'. Pedunculate Oak timber was also used to make furniture and floorboards, offcuts making extremely good firewood. In ecological terms, the importance of Pedunculate Oak to native wildlife cannot be overstated. It supports invertebrate life in abundance, the larvae of several hundred moth species feeding on its leaves, for example. Gall-forming insects (mainly Hymenoptera) are also associated with it and even in death it supports life in the form of wood-boring beetle larvae and fungi.

Purple Hairstreak butterfly, a species tied to mature oaks.

Pedunculate Oak bark

Pedunculate Oak, summer

Pedunculate Oak, winter

Pedunculate Oak acorns

Pedunculate Oak acorns

Pedunculate Oak flowers

Pedunculate Oak Lammas growth

Sessile Oak *Quercus petraea* (Fagaceae) 40m (43m)

Sessile Oak leaf

Sturdy deciduous tree with a domed shape. BARK Grey-brown with deep vertical fissures. BRANCHES Relatively straight, radiating around a longer and more upright bole than Pedunculate Oak. Buds orange-brown with long white hairs. LEAVES Lobed, flattened, dark green and hairless above, paler below with hairs along the veins; on yellow stalks, 1–2.5cm long, and lacking auricles at the base, distinguishing them from those of Pedunculate Oak. REPRODUCTIVE PARTS Drooping green male catkins appear in May and fall off as leaves open fully. Acorns long and egg-shaped, stalkless, sitting directly on the twig in small clusters. STATUS AND DISTRIBUTION Common and widespread in western parts of Britain, in hilly areas on poor soils. COMMENTS Once heavily coppiced for fuel and bark for tanning, but now valued more for its importance to the landscape and the wildlife it supports. Sessile Oak woodlands are often rich in epiphytes. Hole-nesting birds, notably the Pied Flycatcher, are associated with them, feeding on the abundant insect life that Sessile Oak foliage supports.

Sessile Oak acorn

Downy Oak *Quercus pubescens* (Fagaceae) 24m

Downy Oak leaf

Similar to Pedunculate Oak, forming a large, sturdy tree under good growing conditions. BARK Deep grey, grooved with numerous deep fissures and small plates or rough scales. BRANCHES Twigs and buds covered with greyish downy hairs, buds looking more orange-brown beneath the down. LEAVES Smaller than Pedunculate Oak, to 13cm long and 6cm wide, with shallower, forward-pointing lobes and very hairy petioles. Young leaves densely downy at first but becoming smoother and grey-green above when mature. REPRODUCTIVE PARTS Catkins appearing in late May; acorns forming in early autumn. Acorns sessile, borne in stalkless shallow cups about 1.5cm deep, and covered in closely packed downy scales. STATUS AND DISTRIBUTION Native of Europe, occasionally planted in Britain.

Pyrenean Oak *Quercus pyrenaica* (Fagaceae) 15m

Pyrenean Oak leaf

Slender, more open crown than most other oaks. BARK Rough and scaly: a good identification feature. LEAVES Deeply lobed, to 20cm long, with petioles about 2cm long; often on pendulous shoots, and covered with soft grey, downy hairs at first, but becoming smooth above with maturity. REPRODUCTIVE PARTS Male catkins, conspicuously long and yellow, in June and July, often after other oaks have finished flowering. In good years they can be abundant and make a brief but colourful display. Acorns about twice the length of the cup, which is covered in blunt overlapping scales. STATUS AND DISTRIBUTION Native to Iberia, N Italy and Morocco, occasionally planted in Britain. COMMENTS Can be cultivated by grafting shoots onto 2m stocks of Pedunculate Oak.

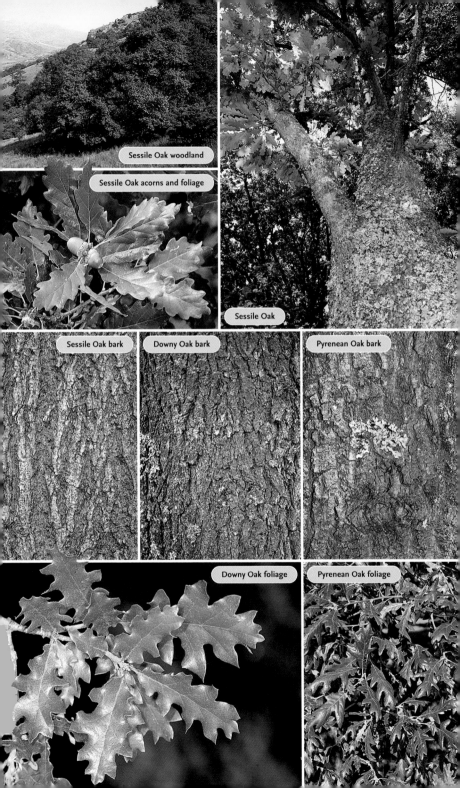

Sessile Oak woodland

Sessile Oak acorns and foliage

Sessile Oak

Sessile Oak bark

Downy Oak bark

Pyrenean Oak bark

Downy Oak foliage

Pyrenean Oak foliage

Cork Oak leaf

Cork Oak *Quercus suber* (Fagaceae) 17m

Medium-sized evergreen oak forming a rounded tree. BARK Thick, pale greyish-brown with deep fissures and ridges if left to mature, and a soft corky texture.

Cork Oak acorn

BRANCHES Numerous, large and twisted, arising low down on bole; in very old trees some branches may trail on ground. LEAVES Like holly leaves, with spiny tips to shallow lobes; to 7cm long, on 1cm petioles. Mature leaves dark green and smooth above, but paler, almost grey and downy below. REPRODUCTIVE PARTS Acorns 2–3cm long, egg-shaped, in cups covered with scales. STATUS AND DISTRIBUTION Native of Mediterranean region, introduced to the British Isles and grown for ornament as far N as Scotland. COMMENTS Abroad, especially in Spain and Portugal, the bark is often regularly stripped to supply corks for wine bottles. After stripping, the trunk is red but the cambium and inner tissues are unharmed and the bark regrows, to be harvested again after a few years.

Turkey Oak *Quercus cerris* (Fagaceae) 38m

Deciduous, broadly conical oak, becoming more spreading and domed with age. BARK Thick, grey-brown, becoming fissured and forming regular, squarish plates in older trees. BRANCHES Appearing swollen near base and spreading upwards. Buds covered with long hairs. LEAVES To 10–12cm long, deeply lobed with up to 10 lobes or large teeth, on slightly downy petioles 1–2cm long. Upper leaf surface feels rough and is deep green, lower surface downy when new and greyish.

Turkey Oak leaf

Turkey Oak acorn

REPRODUCTIVE PARTS Catkins appearing in May–June. Acorns ripening in late summer; partly encased in a deep cup covered in long outward-pointing scales. STATUS AND DISTRIBUTION Native of S Europe, introduced to Britain by J. Lucombe of Exeter in 1735; now widely planted in parks and gardens and sometimes occurring in woodlands. COMMENTS A fast-growing tree, seemingly tolerant of different soil types and atmospheric pollution.

Lucombe Oak *Quercus* × *hispanica* 'Lucombeana' (Fagaceae) 35m

Tall evergreen hybrid between Cork Oak and Turkey Oak. BARK Variable, some specimens similar to Cork Oak, and others having a smoother, darker bark. LEAVES Long, glossy and toothed, remaining on the tree throughout all but the hardest winters. Some of the earliest trees, dating from the original hybridisation, lose a large proportion of their leaves; later crosses have a denser crown. REPRODUCTIVE PARTS Male catkins in early summer; acorns in autumn in small scaly cups. STATUS AND DISTRIBUTION Hybrid originated in Exeter, Devon, in the 18th century, and was named after Lucombe's nursery. It is still most common in parks and gardens around Exeter, especially near the sea. However, it may also be found in mature parks and gardens in sheltered regions elsewhere.

Lucombe Oak leaf

Cork Oak bark

Cork Oak foliage

Cork Oak

Turkey Oak acorn

Turkey Oak bark

Turkey Oak foliage

Lucombe Oak flowers and foliage

Hungarian
Oak leaf

Hungarian Oak *Quercus frainetto* (Fagaceae) 30m

Deciduous, fast-growing oak that forms a fine, broadly domed tree. BARK Pale grey and finely fissured, breaking into fine ridges. BRANCHES Largest ones long and straight, emerging from a sturdy bole and terminating in finely downy greyish-green or brownish twigs. LEAVES Large, deeply lobed, to 25cm long and 14cm wide. REPRODUCTIVE PARTS Pendulous yellow catkins in May and early June; acorns in cups about 1.2cm deep covered in downy, blunt, overlapping scales. STATUS AND DISTRIBUTION Native to Balkans, central Europe and S Italy. Planted elsewhere for its splendid appearance when mature. COMMENTS Sometimes grafted onto the stock of Pedunculate Oak.

Hungarian
Oak acorn

Evergreen
Oak
leaf

Evergreen
Oak acorn

Evergreen (Holm) Oak

Quercus ilex (Fagaceae) 28m

Broadly domed tree; crown is often very dense and twiggy. BARK Very dark with shallow fissures, eventually cracking to form squarish scales. BRANCHES Appearing from low down on bole. Young shoots covered with white down. LEAVES Variable: usually ovate to oblong with a pointed tip and a rounded base on mature trees, but more like holly leaves on a young tree. Dark and glossy above, paler and downy below with raised veins; on hairy petioles 1–2cm long. REPRODUCTIVE PARTS Male catkins appear in spring, their golden colour contrasting with silvery new leaves and darker twigs. Acorns, to 2cm long, sit deeply in cups covered with rows of small hairy scales. STATUS AND DISTRIBUTION Native of S Europe, planted here mainly in mild areas and as a shelter-belt tree in coastal areas, to protect more tender species from winds and salt spray. Naturalised occasionally.

Golden Oak of
Cyprus leaf
underside (*left*) and
upperside (*right*)

SIMILAR TREE

Golden Oak of Cyprus *Q. alnifolia* (8m) A small, shrubby evergreen oak with numerous branches, a short bole and dark grey bark pitted with orange-brown lenticels. Leaves 5cm long, leathery, with a toothed margin, a smooth, dark glossy green upper surface, and distinctive golden felt below. Male catkins yellowish green and pendulous, female catkins smaller and inconspicuous. Acorns up to 3cm long, sitting in a small scaly cup. A native of the mountains of Cyprus, seen in the British Isles only in specialist collections.

Kermes
Oak
leaf

Kermes Oak *Quercus coccifera* (Fagaceae) 5m

Small evergreen oak, often just a dense, much-branched shrub. BARK Greyish and smooth at first, finely patterned in older trees. BRANCHES Young twigs yellowish with branched hairs, but becoming hairless with maturity. LEAVES Tough and holly-like, dark green above and a little paler below, to 4cm long with pronounced spines; petiole short or almost absent. REPRODUCTIVE PARTS Small acorns, to 1.5cm long, sit in a shallow cup protected by strong spiny scales. They take 2 years to mature, so trees always have some acorns on them. STATUS AND DISTRIBUTION Widespread around the Mediterranean. Grown elsewhere for its intriguing foliage; not hardy, so rare in Britain.

Hungarian Oak bark

Hungarian Oak foliage

Hungarian Oak

Evergreen Oak foliage

Evergreen Oak flowers and foliage

Golden Oak of Cyprus foliage

Evergreen Oak acorn

Kermes Oak foliage

Kermes Oak acorns and foliage

Red Oak *Quercus rubra* (Fagaceae) 35m

Broadly conical tree. BARK Pale silvery grey, sometimes brownish, and mostly smooth; fissured with age. LEAVES Large, usually 10–20cm long; deeply lobed, with smaller teeth terminating in fine hairs at tips of lobes. Green above and paler matt green below during growing season; turning red or brown in autumn. Young trees produce the finest red colourings. REPRODUCTIVE PARTS Pendulous male catkins in spring as leaves open, turning tree golden yellow. Acorns rounded, in a neat scaly cup. STATUS AND DISTRIBUTION Native of North America, planted in Britain for autumn colours; naturalised occasionally.

Red Oak acorn

Red Oak leaf

Pin Oak *Quercus palustris* (Fagaceae) 26m

Broadly conical deciduous tree with a short bole. BARK Smooth and greybrown. BRANCHES Numerous, mostly ascending. LEAVES Distinctive: to 12cm long and deeply lobed with bristles at tips of pointed lobes. In summer, leaves are glossy green on both surfaces, palest below; tufts of brownish hairs in vein axils. REPRODUCTIVE PARTS Male catkins pendulous, yellowish, opening in early summer. Acorns, to 1.5cm long, partially enclosed in shallow scaly cup. STATUS AND DISTRIBUTION Native of E North America. Introduced to Britain for ornament.

Pin Oak leaf

Scarlet Oak acorn

Scarlet Oak *Quercus coccinea* (Fagaceae) 28m

Rather slender, domed tree. BARK Dark greyish brown, smooth in young trees, ridged with maturity. BRANCHES Slender and spreading. LEAVES 15cm long, even more deeply lobed than Pin Oak but less strongly bristle-tipped. In summer, glossy green above and paler below with small hair-tufts in vein axils below. Turning brilliant red in autumn, especially in cultivar *Q. coccinea* 'Splendens'. REPRODUCTIVE PARTS Acorns, to 2.5cm long, are rounded, half-enclosed in a slightly glossy cup. STATUS AND DISTRIBUTION Native of E North America. Planted in Britain for its brilliant autumn colours.

Scarlet Oak leaf

Mirbeck's Oak leaf

Mirbeck's Oak *Quercus canariensis* (Fagaceae) 25m

Domed, columnar tree. BARK Thick, dark grey and furrowed. LEAVES Ovate to elliptic, to 15cm long, with up to 12 lobes. Young leaves hairy and reddish, maturing darker green and smooth. Some turn yellow and fall in autumn, others remain through winter. REPRODUCTIVE PARTS Male catkins yellowish green and pendulous; female catkins small. Acorns, to 2.5cm long, ovate, one-third hidden in scaly cup. STATUS AND DISTRIBUTION Native of N Africa and SW Europe, planted in Britain occasionally.

Mirbeck's Oak acorn

Water Oak *Quercus nigra* (Fagaceae) 18m

Domed tree. BARK Purplish grey. BRANCHES Spreading. LEAVES Dark green, glossy, hairless, with irregular lobes, broadest near blunt tip; retained into winter. REPRODUCTIVE PARTS Domed acorns in shallow cups. STATUS AND DISTRIBUTION Native of E USA, planted in Britain occasionally.

Water Oak leaf

Red Oak 'Aurea' autumn colours

Scarlet Oak autumn colours

Red Oak bark

Pin Oak autumn colours

Pin Oak bark

Scarlet Oak bark

Mirbeck's Oak acorns and foliage

Mirbeck's Oak bark

Water Oak foliage

Water Oak bark

ELMS *ULMUS* (FAMILY ULMACEAE)

A genus of mostly large deciduous trees with small flowers that open before the leaves, except for a few that flower in autumn. Leaves are asymmetric at the base. Many species propagate freely from root suckers.

Wych Elm *Ulmus glabra* (Ulmaceae) 40m

Wych Elm leaf

Large, often spreading tree, frequently with several prominent trunks arising from a stout bole. Rarely produces suckers like other elms, so reproduces only by seed. BARK Smooth and greyish in younger trees, becoming browner with deep, mostly vertical cracks and ridges with age. BRANCHES Main ones spreading, sometimes almost horizontal. Youngest twigs thick, reddish brown and covered with short stiff hairs; older twigs smoother and greyer. In winter, buds are reddish brown and hairy, oval with blunt-pointed tips. LEAVES Rounded or oval, to 18cm long, with long tapering point at tip. Base of leaf unequal: a good pointer to all the elms. Long side of leaf base extends beyond petiole (which is 2–5mm long) to the twig. Leaves feel rough; upper surface hairy, and lower surface with softer, sparser hairs. REPRODUCTIVE PARTS Flowers sessile with purple anthers, opening before leaves in February and March, high up on tree so inconspicuous. Fruits about 2cm long, on a short stalk, and papery. STATUS AND DISTRIBUTION Native of much of Europe, including Britain and Ireland, occurring in woods and especially hedgerows, often near flowing water. An attractive feature of many riversides in the north of England. COMMENTS Susceptible to Dutch elm disease, so much reduced in numbers. Slow to recover as it does not sucker like other elms.

English Elm *Ulmus procera* (Ulmaceae) 36m

High-domed and lofty. BARK Dark brown, grooved with small squarish plates. BRANCHES Main ones large and ascending. Twigs thick, reddish and densely hairy. Winter buds 3mm long, ovoid, pointed and hairy (use a hand-lens). LEAVES Rounded or slightly oval with short tapering tip; base unequal, longest side does not reach beyond petiole to twig. Leaf rough to touch; petiole (1–5mm long) and midrib finely downy. REPRODUCTIVE PARTS Flowers with dark red anthers, opening before leaves in February and March. Ripe fruits (rarely produced) up to 1.5cm long, papery, and very short-stalked. STATUS AND DISTRIBUTION Native of S and E Europe, in hedgerows and woodland edges. Doubtfully native to Britain, except perhaps in S England, but present for millennia.

English Elm leaf

COMMENTS Probably brought to Britain by early European colonisers; certainly introduced to Ireland and N Britain. Sadly, no longer a countryside feature because of Dutch elm disease. Formerly, its leaves were fed to cattle. Its durable timber was used for furniture, floorboards and coffins. Some authorities classify it as *U. minor* ssp. *vulgaris*.

Wych Elm fruits

Wych Elm bark

English Elm bark

Wych Elm foliage

English Elm foliage

English Elm fruits

English Elm

Ulmus minor is a variable tree and its classification is extremely confusing. Formerly, several regional subspecies were recognised and given English names. However, I have adopted the classification system suggested by Max Coleman in *British Wildlife* Vol. 13, No. 6. Smooth-leaved Elm *U. minor* is recognised as a native species. Cornish, Jersey and Plot's Elms, while reasonably distinct, are now considered to be clones of *U. minor*, their current distribution influenced to varying degrees by the planting of cuttings; consequently, a horticultural style is adopted for their scientific names.

Smooth-leaved Elm *Ulmus minor* (Ulmaceae) 32m

Domed and spreading tree. BARK Greyish brown, scaly and ridged. BRANCHES Usually ascending, often with pendulous masses of shoots. LEAVES Superficially hornbeam-like, leathery, to 9cm long, oval, pointed at tip, with toothed margins; unequal leaf bases, narrowly tapering on short side, and a short petiole.

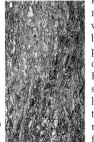

REPRODUCTIVE PARTS Fruits papery. STATUS AND DISTRIBUTION Native to S and SE England; once widespread but range and abundance badly affected by Dutch elm disease. COMMENTS Includes trees previously (and sometimes still) known as *U. carpinifolia* and Coritanian Elm *U. coritana*. Coritanian Elm is often distinct enough to be recognised. It is a spreading tree with stout twigs and rather broad leaves that show a distinct curve towards the side of the leaf where the margin is shortest. The base is markedly unequal, with the long side sometimes forming a lobe. It is restricted to E and SE Britain.

Smooth-leaved Elm leaf (*left*) and bark (*right*)

Cornish Elm leaf

Cornish Elm *Ulmus* 'Stricta' (Ulmaceae) 36m

(Referred to in Stace as *Ulmus minor* ssp. *angustifolia*.) Narrowly conical hedgerow tree with a rather open, spreading crown. BARK Grey-brown and scaly. BRANCHES Relatively few; lowest ones ascending steeply. LEAVES Oval, toothed and relatively small (to 6cm); smooth and leathery above, downy on midrib below. Leaf narrow and almost equal at base, sometimes concave and with a straight midrib. Petiole 1cm long and downy. REPRODUCTIVE PARTS Papery fruits. STATUS AND DISTRIBUTION Restricted mainly to Cornwall, W Devon; more local elsewhere in SW England and in SW Ireland. Much reduced because of Dutch elm disease. A form known as Goodyer's Elm occurs in S Hampshire and is sometimes afforded species status (*U. angustifolia*).

Jersey Elm *Ulmus* 'Sarniensis' (Ulmaceae) 20m

(Referred to in Stace as *Ulmus minor* ssp. *sarniensis*.) Similar to Cornish Elm but separable by using characters as well as geographical range. Has a tall, straight trunk that extends to top of tree and a conical shape overall in maturity. BRANCHES Numerous; spreading or only slightly ascending. LEAVES To 4cm long, rather neatly narrowly ovate and almost equal at the base. REPRODUCTIVE PARTS Papery fruits. STATUS AND DISTRIBUTION Widespread on Guernsey, probably introduced to other Channel Islands, and occasionally planted in mainland Britain too as a street tree, for example, in Edinburgh.

Jersey Elm leaf

Smooth-leaved Elm foliage

Cornish Elm foliage

Jersey Elm foliage

Smooth-leaved Elm hedgerow

Cornish Elm hedgerow

Plot's
Elm leaf

Plot's Elm *Ulmus* 'Plotii' (Ulmaceae) 25m

(Referred to in Stace as *Ulmus plotii*.) Reasonably distinctive tree with a narrow, upright crown, an arching leading shoot and overall a rather shaggy appearance in maturity. BARK Greyish brown and scaly. BRANCHES Rather slender with long pendulous twigs. LEAVES To 4cm long, narrow (much more so than English Elm), widest in the middle with a straight midrib and pointed tip, and an almost equal base; upper surface rough to touch. REPRODUCTIVE PARTS Papery fruits. STATUS AND DISTRIBUTION A hedgerow and field-margin tree, more or less confined to the English East Midlands. Much reduced (large trees in particular) by Dutch elm disease. Favours damp ground.

Dutch
Elm
fruit

Dutch
Elm leaf

Dutch Elm *Ulmus* × *hollandica* (Ulmaceae) 30m

Tall and rather straggly hybrid tree. BARK Brown, cracking into small, shallow plates. BRANCHES Higher branches longer than ones lower down, and spreading. LEAVES Oval, toothed, to 15cm long, sometimes buckled. Leaf base only slightly unequal. REPRODUCTIVE PARTS Papery fruits. STATUS AND DISTRIBUTION Hybrid (probably naturally occurring), whose parents are *U. glabra* and *U. minor*. Has a scattered range across S England and SW Wales and is found in hedgerows in lowland districts. COMMENT Has a degree of natural resistance to Dutch elm disease and this is enhanced in some cultivars, notably 'Groeneveld'.

Huntingdon Elm *Ulmus* × *hollandica* 'Vegeta' (Ulmaceae) 30m

(Referred to in Stace as *Ulmus* × *vegeta*.) Widely spreading tree with a domed crown. BARK Greyish and broken into regular ridges. BRANCHES Main branches long, straight and upright to spreading. LEAVES Ovate to elliptical with a pointed tip and toothed margins; base markedly unequal and upper surface smooth. Rather similar to leaves of Wych Elm, but petiole more than 5mm long. REPRODUCTIVE PARTS Papery fruits. STATUS AND DISTRIBUTION A clone of Dutch Elm, formerly considered to be a naturally occurring hybrid, found in East Anglia and central England. It is widely planted. COMMENTS Has a degree of resistance to Dutch elm disease.

Huntingdon
Elm leaf

Other Hybrid Elms *Ulmus*; many different hybrids (Ulmaceae) 35m

Narrowly columnar trees with rather dense foliage exist. BARK Brown, cracking into small, square plates. BRANCHES Upright, straight and spreading at shallow angles. LEAVES Ovate to elliptical, dark green and shiny with a pointed tip and toothed margins; base almost equal and upper surface smooth. REPRODUCTIVE PARTS Papery fruits. STATUS AND DISTRIBUTION Widely planted. COMMENTS Artificial hybrids with a complex parentage that includes *U.* × *hollandica*. Frequent cultivars include 'Lobel' with oval leaves and 'Dodoens' with leaves that have a very long, pointed tip.

Hybrid
Elm
'Lobel' leaf

Hybrid Elm
'Dodoens'
leaf

Plot's Elm foliage

Dutch Elm bark

Dutch Elm fruits

Dutch Elm

Dutch Elm foliage

Huntingdon Elm foliage

Hybrid Elm 'Lobel' foliage

Hybrid Elm 'Dodoens' foliage

European White Elm leaf

European White Elm *Ulmus laevis* (Ulmaceae) 20m
Broadly spreading tree with an open crown. BARK Grey and smooth when young, deeply furrowed with age. BRANCHES Twigs reddish brown and softly downy, but becoming smooth with age. LEAVES To 13cm long, with markedly unequal bases and toothed margins. Leaf veins paired, and longer side has 2–3 more veins than the other. Upper leaf surface usually smooth but underside normally with grey down. REPRODUCTIVE PARTS Flowers in long-stalked clusters. Fruits winged and papery, with a fringe of hairs; in pendulous clusters. STATUS AND DISTRIBUTION Native of mainland Europe, possibly native in Britain in the past but now probably extinct. Sometimes grown in collections.

Japanese Elm leaf

Chinese Elm leaf

Japanese Elm *Ulmus japonica* (Ulmaceae) 8m
Spreading, low-growing tree. BARK Grey-brown, scaly and ridged. BRANCHES Upright and spreading with downy shoots. LEAVES Narrow-ovate, to 10cm long, dark green, rough above, downy below; leaf bases unequal. REPRODUCTIVE PARTS Papery fruits. STATUS AND DISTRIBUTION Native of Japan, planted in Britain partly for its resistance to Dutch elm disease.

SIMILAR TREE
Chinese Elm *U. parvifolia* (15m) Crown domed; leaves oval, dark green, to 6cm long; bases almost equal and teeth blunt. Native of E Asia, sometimes planted.

ZELKOVA (FAMILY ULMACEAE)
Includes 5 species, closely related to the Elms, from E Mediterranean, the Caucasus, China and Japan. Only one becomes a large tree. Susceptible to Dutch elm disease.

Caucasian Elm leaf

Keaki leaf

Caucasian Elm *Zelkova carpinifolia* (Ulmaceae) 31m
Dense, multi-stemmed crown composed of numerous almost upright branches. Bole, to 3m, is heavily ridged. BARK Greyish and flaking; falling away in rounded scales to expose orange patches. BRANCHES Youngest twigs greenish with whitish down. LEAVES To 10cm long, oval and pointed with rounded teeth and 6–12 pairs of veins. Upper surface dark green and slightly hairy, lower surface slightly paler with hairs on either side of veins. Petiole very short, to 2mm long. REPRODUCTIVE PARTS Male flowers, in April, are sessile clusters of yellow-green stamens arising from older, leafless part of twig. Female flowers solitary, in axils of last few leaves on shoot. Fruits spherical, to 5mm across and slightly 4-winged. STATUS AND DISTRIBUTION Native of the Caucasus, grown for ornament elsewhere. COMMENTS Noted for its autumn colours. Mature trees produce suckers and can spread like a hedgerow elm.

SIMILAR TREE
Keaki *Z. serrata* (26m) Young twigs hairy at first, becoming smoother with age. Leaves more markedly toothed and smooth below. Fruits smooth and rounded.

European White Elm bark

European White Elm autumn foliage

Japanese Elm foliage

Chinese Elm fruits and foliage

Caucasian Elm

Caucasian Elm bark

Caucasian Elm foliage

Keaki foliage

Black Mulberry *Morus nigra* (Moraceae) 13m

Gnarled bole and dense, twisting branches and twigs make even a young tree look ancient. Crown may be broader than tree is tall. BARK Dark orange-brown, fissured and peeling. Downy shoots release milky juice if snapped. LEAVES To 20cm long, oval with heart-shaped base, toothed margin and pointed tip. Petiole hairy, to 2.5cm long. REPRODUCTIVE PARTS Flower spikes on short downy stalks in May; yellowish-green male flowers about 2.5cm long; females about 1–1.25cm long, giving rise to a hard raspberry-like fruit, acidic until fully ripened, when wine-red or purple. STATUS AND DISTRIBUTION Native of Asia, long cultivated elsewhere. In Britain, found mainly in the south, in sheltered gardens.

Black Mulberry leaf

White Mulberry *Morus alba* (Moraceae) 15m

White Mulberry leaf

Deciduous tree with a narrow rounded crown on a broad bole, to 2m across. BARK Heavily ridged and grey, sometimes tinged pinkish. BRANCHES Shoots thin, with fine hairs at first; buds minute, brown and pointed. LEAVES To 18cm long, oval to rounded with a heart-shaped base and a hairy, grooved petiole up to 2.5cm long. They feel thin and smooth, and have a toothed margin, with downy hairs on veins on underside. REPRODUCTIVE PARTS Female flowers stalked, spike-like and yellowish. Male flowers on slightly longer spikes; whitish with prominent anthers. Fruit comprises a cluster of drupes; white or pink at first, ripening purple. STATUS AND DISTRIBUTION Native of E Asia, grown in Britain occasionally. COMMENT The leaves are the food plant for silkworms.

Fig leaf

Fig *Ficus carica* (Moraceae) 5m

Deciduous tree with distinctive fruits and leaves. BARK Pale grey, smooth, sometimes with finer lines. BRANCHES Thick, forming a spreading domed crown. LEAVES Alternate, to 20cm long, on a 5–10cm petiole; deeply lobed, usually in 3 segments, sometimes 5. They feel rough and leathery, with prominent veins on underside. REPRODUCTIVE PARTS Flowers hidden, produced inside pear-like, fleshy receptacle that is almost closed at apex. This ripens in second year into the familiar fleshy, sweet-tasting fig. STATUS AND DISTRIBUTION Native to SW Asia, possibly also S and E Europe. Long cultivated in Britain, thriving in walled gardens. COMMENTS Cultivars have been developed and these are more likely to be seen in gardens. The fruits are eaten either fresh or dried.

Fig fruit

Barberry *Berberis vulgaris* (Berberidaceae) 2m

Barberry leaf

Small deciduous shrub with grooved twigs and 3-forked prickles. REPRODUCTIVE PARTS Flower small, yellow, in hanging clusters in late spring. Fruits ovoid, reddish berries. LEAVES Sharp-toothed, oval; borne in tufts from axils of prickles. STATUS AND DISTRIBUTION Scarce native in Britain but also planted and naturalised. Found in hedgerows and scrub, mainly on calcareous soils.

Barberry fruit

Black Mulberry bark

Black Mulberry foliage

Black Mulberry unripe fruits

White Mulberry fruits and foliage

Black Mulberry ripe fruits

Barberry flowers and foliage

Fig foliage

Fig fruits

Chilean Firebush *Embothrium coccineum* (Proteaceae) 12m

Small, spreading and untidy-looking evergreen. BARK Purple-grey and flaking. BRANCHES With slightly pendulous shoots. LEAVES Lanceolate, to 22cm long. REPRODUCTIVE PARTS Best known for its clusters of striking red flowers, to 10cm long, produced in May; swollen at tip, which, when open, divides into 4 segments. Fruits are grooved capsules that retain 3cm-long style at tip. STATUS AND DISTRIBUTION Native of Chile and Argentina, planted elsewhere for its colourful flowers.

Katsura
Tree leaf

Katsura Tree *Cercidiphyllum japonicum* (Cercidiphyllaceae) 25m

Conical-crowned deciduous tree, sometimes with a single bole, more often with several main stems. BARK Vertically fissured and peeling. LEAVES In opposite pairs, to 8cm long, rounded, with pointed tips and heart-shaped bases. Pink at first, turning green in summer, then red in autumn. REPRODUCTIVE PARTS Flowers in leaf nodes in April. Male flowers are small clusters of reddish stamens, female flowers are darker red clusters of styles. Fruits are claw-like bunches of 5cm-long pods that change from grey, through green, to brown. STATUS AND DISTRIBUTION Native of Japan, grown in Britain and Ireland for ornament.

Katsura Tree bark

TULIP TREES *LIRIODENDRON* (FAMILY MAGNOLIACEAE)

Only 2 species occur, relics of the pre-Ice Age flora, surviving in North America and China.

Tulip
Tree
leaf

Tulip Tree *Liriodendron tulipifera* (Magnoliaceae) 45m

Impressive deciduous tree. Despite the promise of its name, foliage is more attractive than flowers. BARK Pale grey. BRANCHES Often twisted. LEAVES Strikingly shaped, to 20cm long and 4-lobed with a terminal notch, looking as though they have been cut out with scissors; fresh green through summer, turning bright gold in autumn. Leaves smooth and hairless, on a slender petiole 5–10cm long. REPRODUCTIVE PARTS Flowers superficially tulip-like, giving the tree its name. Cup-shaped at first, and inconspicuous as perianth segments are greenish and blend in with leaves. Later, flowers open more fully, revealing rings of yellowish stamens surrounding paler ovaries. Often produced high up in the middle of dense foliage, and not until the tree is at least 25 years old and quite sizeable. Conical fruits, to 8.5cm long, are composed of numerous scale-like overlapping carpels. STATUS AND DISTRIBUTION Native of E USA. Introduced to Europe in the 17th century and commonly planted in gardens and parks.

Tulip Tree bark

Chinese Tulip Tree leaf

SIMILAR TREE

Chinese Tulip Tree *Liriodendron chinense* (25m) Similar to Tulip Tree in most respects but leaves more narrowly waisted and terminal 'cut' often less indented. Native to E Asia and planted in Britain occasionally.

Chilean Firebush flowers and foliage

Katsura Tree autumn colour

Katsura Tree foliage

Tulip Tree

Tulip Tree autumn colour

Chinese Tulip Tree foliage

Tulip Tree flower

MAGNOLIAS *MAGNOLIA* (FAMILY MAGNOLIACEAE)

A genus of about 35 species, many with beautiful showy but primitive flowers. Numerous cultivars exist and these are popular garden trees.

Southern Evergreen Magnolia leaf upperside

Southern Evergreen Magnolia leaf underside

Southern Evergreen Magnolia (Bull Bay)

Magnolia grandiflora (Magnoliaceae) 30m

A large, spreading evergreen tree with a broadly conical crown. BARK Smooth, dull grey. BRANCHES Large, the youngest shoots covered with thick down and terminating in red-tipped buds. LEAVES Elliptical, to 16cm long and 9cm wide with a smooth or sometimes wavy margin. Upper surface shiny, dark green, and underside rust-coloured and downy, as is the 2.5cm-long petiole. REPRODUCTIVE PARTS Striking flowers, composed of 6 white petal-like segments, borne at tips of shoots; conical in bud, later opening out to a spreading cup-shape, to 25cm across. Fruit conical, to 6cm long, composed of scale-like carpels on a single orange stalk. Flowers from midsummer to late autumn. STATUS AND DISTRIBUTION Native of SE USA, introduced to Europe in 18th century. Popular in gardens, and does well if grown against a wall. In more sheltered areas it will form a splendid free-standing tree. COMMENTS Many other magnolia species and cultivars are found in cultivation and the following are among the most popular in gardens:

Campbell's Magnolia leaf

Campbell's Magnolia *Magnolia campbellii* (Magnoliaceae) 20m

Much-branched and widely spreading tree, or large bush. BARK Grey, slightly fissured. BRANCHES Long, mainly level. LEAVES Oval, pointed at the tip. REPRODUCTIVE PARTS Flowers, to 30cm across, comprising numerous pink tepals. STATUS AND DISTRIBUTION Native of E Asia, grown in gardens elsewhere for its wonderful flowers. COMMENTS Used as a parent stock in many hybrid tree magnolia cultivars, although in many forms it is not possible to determine the precise parentage.

Star Magnolia leaf

Star Magnolia

Magnolia stellata (Magnoliaceae) 9m

Usually forms a much-branched bush rather than a tree. LEAVES Narrowly oval. REPRODUCTIVE PARTS Flowers comprising numerous white tepals, arranged in a star-like fashion. STATUS AND DISTRIBUTION Native of Japan and widely planted in Britain.

Saucer Magnolia

Magnolia × soulangeana (Magnoliaceae) 12m

Dense bush or untidy, low, spreading tree. BARK Grey. BRANCHES Much divided and untidy; shoots carrying silky buds. LEAVES Dark green, ovate and pointed at the tip. REPRODUCTIVE PARTS Upright flowers, the 9 pale tepals flushed pinkish orange or purple at the base. STATUS AND DISTRIBUTION Widely planted hybrid and a contributor to many other hybrid cultivars.

Saucer Magnolia leaf

Southern Evergreen Magnolia flower and foliage

Campbell's Magnolia flowers

Star Magnolia flower and foliage

Magnolia 'Loeberi' – a hybrid between *M. stellata* and *M. kobus*

Saucer Magnolia flower

Saucer Magnolia flower

Magnolia 'Iolanthe' flower

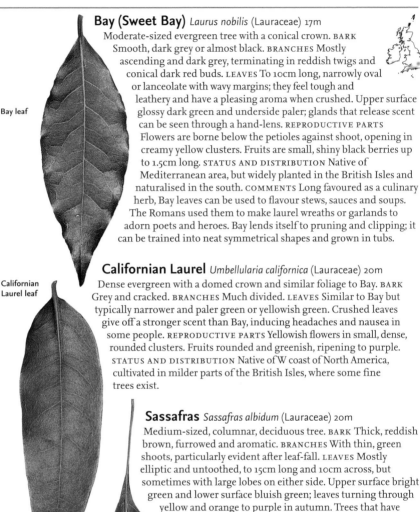

Bay leaf

Californian
Laurel leaf

Sassafras leaf

Bay (Sweet Bay) *Laurus nobilis* (Lauraceae) 17m

Moderate-sized evergreen tree with a conical crown. BARK Smooth, dark grey or almost black. BRANCHES Mostly ascending and dark grey, terminating in reddish twigs and conical dark red buds. LEAVES To 10cm long, narrowly oval or lanceolate with wavy margins; they feel tough and leathery and have a pleasing aroma when crushed. Upper surface glossy dark green and underside paler; glands that release scent can be seen through a hand-lens. REPRODUCTIVE PARTS Flowers are borne below the petioles against shoot, opening in creamy yellow clusters. Fruits are small, shiny black berries up to 1.5cm long. STATUS AND DISTRIBUTION Native of Mediterranean area, but widely planted in the British Isles and naturalised in the south. COMMENTS Long favoured as a culinary herb, Bay leaves can be used to flavour stews, sauces and soups. The Romans used them to make laurel wreaths or garlands to adorn poets and heroes. Bay lends itself to pruning and clipping; it can be trained into neat symmetrical shapes and grown in tubs.

Californian Laurel *Umbellularia californica* (Lauraceae) 20m

Dense evergreen with a domed crown and similar foliage to Bay. BARK Grey and cracked. BRANCHES Much divided. LEAVES Similar to Bay but typically narrower and paler green or yellowish green. Crushed leaves give off a stronger scent than Bay, inducing headaches and nausea in some people. REPRODUCTIVE PARTS Yellowish flowers in small, dense, rounded clusters. Fruits rounded and greenish, ripening to purple. STATUS AND DISTRIBUTION Native of W coast of North America, cultivated in milder parts of the British Isles, where some fine trees exist.

Sassafras *Sassafras albidum* (Lauraceae) 20m

Medium-sized, columnar, deciduous tree. BARK Thick, reddish brown, furrowed and aromatic. BRANCHES With thin, green shoots, particularly evident after leaf-fall. LEAVES Mostly elliptic and untoothed, to 15cm long and 10cm across, but sometimes with large lobes on either side. Upper surface bright green and lower surface bluish green; leaves turning through yellow and orange to purple in autumn. Trees that have spread through suckers can form large clumps that produce a brilliant autumn display. Crushed leaves have a pleasing smell and to some they taste of orange and vanilla. REPRODUCTIVE PARTS Male and female flowers very small, greenish yellow, without petals, in small clusters on separate plants and opening in the spring. Fruit an ovoid berry, about 1cm long, ripening to dark blue. STATUS AND DISTRIBUTION Common native tree of E North America, growing in woods and thickets, and used as a raw ingredient for root beer and tea. Seen in Britain and Ireland in arboreta and well-established gardens.

Bay flowers and foliage

Californian Laurel flowers and foliage

Californian Laurel foliage

Sassafras foliage

Sassafras bark

Sweet Gum *Liquidambar styraciflua* (Hamamelidaceae) 28m

Sweet Gum
fruit

A large tree with attractive foliage. BARK Greyish brown with scaly ridges. BRANCHES Twisting and spreading to upcurved. LEAVES Sharply lobed with a toothed margin. They are alternate and give off a resinous scent when crushed, unlike maple leaves, which they resemble. REPRODUCTIVE PARTS Flowers globose; fruits spiny and pendulous, 2.5–4cm across, resembling those of a plane. STATUS AND DISTRIBUTION A widespread and common native tree of the SE USA as far S as Central America. Familiar as a colourful autumn tree in many British parks and gardens. Also imported as timber called satinwood.

Sweet Gum
leaf in autumn

Persian Ironwood *Parrotia persica*
(Hamamelidaceae) 12m

A small, spreading deciduous tree with a short bole. BARK Smooth, peels away in flakes, leaving attractive coloured patches; older trees have a pattern of pink, brown and yellow. BRANCHES Mostly level. Young twigs hairy, terminating in blackish hairy buds. LEAVES To 7.5cm long, oval with a slightly tapering tip and a rounded base; margins wavy or sometimes slightly toothed. Glossy green above and appearing slightly crushed or crinkled; underside slightly hairy. REPRODUCTIVE PARTS Flowers appear before the leaves in early spring, in short-stalked clusters, reddish and inconspicuous. Fruits are dry capsules that split open to release small, pointed, shiny brown seeds. STATUS AND DISTRIBUTION Native of the Caucasus and N Iran, introduced to Europe as an ornamental tree.

Persian Ironwood
leaf in summer

Persian Ironwood
leaf in autumn

Witch Hazel *Hamamelis mollis* (Hamamelidaceae) 4m

Rarely more than a small sprawling shrub, but sometimes grows into a small domed tree. BARK Greyish brown. BRANCHES Dense and mostly ascending. LEAVES Resembling Hazel leaves; alternate and mostly oval with pointed tips, a toothed margin and an unequal base.
REPRODUCTIVE PARTS Best known for its winter flowers, produced long before the leaves open and providing a welcome sign of early spring in a winter landscape. They are composed of long yellow, ribbon-like petals and red stamens, and are noticeably sweet-scented. They are often produced prolifically on bare twigs, making the shrub stand out conspicuously. STATUS AND DISTRIBUTION Native of China, introduced to Britain late in the 19th century and now found in parks and gardens and sometimes naturalised in open woodlands.

Witch
Hazel leaf

Japanese
Witch Hazel
leaf

Hybrid Witch
Hazel leaf

SIMILAR TREES

Japanese Witch Hazel *H. japonica* (4m) A spring-flowering shrub with a spreading habit. Flower colour rather subdued but autumn leaves colourful.
Hybrid Witch Hazel *H.* × *intermedia* (4m) A popular hybrid with more showy spring flowers than the above species; many different cultivars exist.

Sweet Gum foliage

Persian Ironwood bark

Sweet Gum

Persian Ironwood flowers and foliage

Witch Hazel flowers

Hybrid Witch Hazel flowers

Hybrid Witch Hazel fruits

Japanese Witch Hazel flowers

Oriental Plane *Platanus orientalis* (Platanaceae) 30m

Oriental Plane leaf

Large deciduous tree with a broad, domed crown. The main trunk is frequently covered with large tuberous burrs. BARK Mostly smooth and pale brown, flaking away to reveal rounded yellow patches. BRANCHES Often spreading. In older specimens branches droop down to the ground. Young shoots yellow-brown and hairy; older twigs greyer. LEAVES Large, up to 18cm in length and width; deeply divided into 5–7 lobes that are themselves notched; central lobe longest. Petiole 5cm long with swollen base enclosing a bud. REPRODUCTIVE PARTS Male flowers, up to 6cm long, composed of 2–7 rounded, yellowish flower heads. Female flowers, up to 8cm long, comprising up to 6 rounded, dark red flower heads; flowers open May–June. As they ripen into fruits, the catkins reach a length of 15cm and the ball-like heads grow to 3cm across; they contain many 1-seeded carpels with long hairs attached to bases. STATUS AND DISTRIBUTION Native of the Balkans, eastwards into Asia. Commonly planted in British parks and gardens, sometimes alongside roads, although it is not suitable for narrow roads because its spreading habit makes it too large.

Oriental Plane 'Digitata' leaf

London Plane *Platanus × hispanica* (× *acerifolia*) (Platanaceae) 44m

Large deciduous tree resulting from a cross between the American and Oriental Planes, and known since the mid-17th century. Main trunk is usually very tall and the crown of an old tree is often spreading. BARK Greyish-brown and thin, flaking away in rounded patches to leave paler, yellowish areas beneath. BRANCHES Often tangled and rather twisted. LEAVES Up to 24cm long and mostly 5-lobed and palmate, but not as deeply divided as those of Oriental Plane; very variable, however, and the degree of lobing may differ greatly. REPRODUCTIVE PARTS Flowers very similar to those of Oriental Plane. Some flower spikes may bear only 1 or 2 flower heads. Fruits also similar to Oriental Plane. STATUS AND DISTRIBUTION A widespread tree in towns, where the peeling bark is a useful way of ridding the tree of soot and dust deposits. Very sturdy and resistant to gales and storms, and also quite disease-resistant. Very much a feature of London streets and squares – hence the name – but also popular in other cities.

Oriental Plane bark

London Plane bark

London Plane leaf

London Plane fruit

Oriental Plane fruits and foliage

Oriental Plane fruits

London Plane fruits and foliage

London Plane foliage

Oriental Plane

London Plane

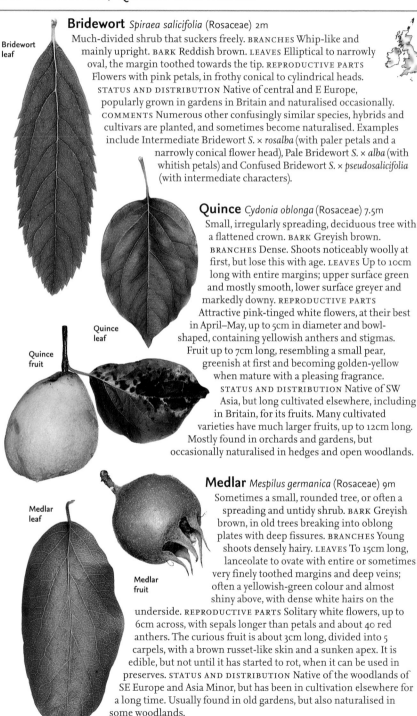

Bridewort *Spiraea salicifolia* (Rosaceae) 2m

Bridewort leaf

Much-divided shrub that suckers freely. BRANCHES Whip-like and mainly upright. BARK Reddish brown. LEAVES Elliptical to narrowly oval, the margin toothed towards the tip. REPRODUCTIVE PARTS Flowers with pink petals, in frothy conical to cylindrical heads. STATUS AND DISTRIBUTION Native of central and E Europe, popularly grown in gardens in Britain and naturalised occasionally. COMMENTS Numerous other confusingly similar species, hybrids and cultivars are planted, and sometimes become naturalised. Examples include Intermediate Bridewort *S.* × *rosalba* (with paler petals and a narrowly conical flower head), Pale Bridewort *S.* × *alba* (with whitish petals) and Confused Bridewort *S.* × *pseudosalicifolia* (with intermediate characters).

Quince *Cydonia oblonga* (Rosaceae) 7.5m

Small, irregularly spreading, deciduous tree with a flattened crown. BARK Greyish brown. BRANCHES Dense. Shoots noticeably woolly at first, but lose this with age. LEAVES Up to 10cm long with entire margins; upper surface green and mostly smooth, lower surface greyer and markedly downy. REPRODUCTIVE PARTS Attractive pink-tinged white flowers, at their best in April–May, up to 5cm in diameter and bowl-shaped, containing yellowish anthers and stigmas. Fruit up to 7cm long, resembling a small pear, greenish at first and becoming golden-yellow when mature with a pleasing fragrance. STATUS AND DISTRIBUTION Native of SW Asia, but long cultivated elsewhere, including in Britain, for its fruits. Many cultivated varieties have much larger fruits, up to 12cm long. Mostly found in orchards and gardens, but occasionally naturalised in hedges and open woodlands.

Quince leaf

Quince fruit

Medlar *Mespilus germanica* (Rosaceae) 9m

Medlar leaf

Sometimes a small, rounded tree, or often a spreading and untidy shrub. BARK Greyish brown, in old trees breaking into oblong plates with deep fissures. BRANCHES Young shoots densely hairy. LEAVES To 15cm long, lanceolate to ovate with entire or sometimes very finely toothed margins and deep veins; often a yellowish-green colour and almost shiny above, with dense white hairs on the underside. REPRODUCTIVE PARTS Solitary white flowers, up to 6cm across, with sepals longer than petals and about 40 red anthers. The curious fruit is about 3cm long, divided into 5 carpels, with a brown russet-like skin and a sunken apex. It is edible, but not until it has started to rot, when it can be used in preserves. STATUS AND DISTRIBUTION Native of the woodlands of SE Europe and Asia Minor, but has been in cultivation elsewhere for a long time. Usually found in old gardens, but also naturalised in some woodlands.

Medlar fruit

Bridewort flowers and foliage

Quince flower

Quince fruit and foliage

Medlar flower

Medlar fruit

PEARS *PYRUS* (FAMILY ROSACEAE)

About 20 species occur in temperate regions of Europe and Asia. Some produce edible fruits and many yield good-quality durable timber that is used for inlay work and turnery. Many are found in cultivated forms and are important commercial species.

Common Pear (Cultivated Pear) *Pyrus communis* (Rosaceae) 20m

A normally upright and slender deciduous tree with a stout bole and a dense framework unless pruned. BARK Dark brown, breaking up into small square plates. BRANCHES Ascending in young trees, but becoming more spreading in older specimens; some branches may bear a few spines. Young twigs reddish brown and sparsely hairy, becoming smoother with age. LEAVES Up to 8cm long, usually oval to elliptic in shape, but always with some variation; the margins have numerous small teeth, and the leaves are smooth and almost glossy when mature. REPRODUCTIVE PARTS Flowers pure white, opening before leaves have fully expanded, typically 2 to 4 weeks earlier than cultivated apples flowering in the same location. A pear orchard is a spectacular sight on a sunny spring day. The fruits may be up to 12cm long, with soft, but slightly gritty, sweet-tasting flesh. STATUS AND DISTRIBUTION A native of W Asia originally, but cultivated for millennia and now widespread across Europe, including Britain and Ireland. Grown traditionally for its highly edible fruits but nowadays also planted simply for its display of flowers (often using sterile hybrids). Many hundreds of cultivars exist and there are numerous hybrids. Today, almost all Common Pears that are grown and sold are grafted onto rootstocks and this can affect their size and appearance. Popular varieties include 'Conference' (good for eating) and 'Concorde' (grown as a dessert pear). Single trees may be seen naturalised in hedgerows and woodlands, sometimes indicating where an old dwelling once existed. COMMENTS Cultivated Pear can be trained into a variety of shapes, by a range of different techniques (e.g. cordon and espalier), in order to aid fruit picking or to make best use of available space. The close-grained timber was, and still is, used in turnery: particularly prized objects include bowls and urns.

Common Pear leaf

Common Pear fruit

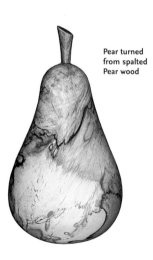

Pear turned from spalted Pear wood

Common Pear bark

Common Pear

Common Pear with fruit

Common Pear fruit and foliage

Common Pear flowers

Plymouth Pear *Pyrus cordata* (Rosaceae) 8m

A small, slender or slightly spreading deciduous tree. BARK Dark brown, breaking up into small square plates. BRANCHES Spiny, with purplish twigs. LEAVES Alternate, oval and up to 5cm long, although they are usually much smaller. Margin finely toothed and leaf downy when young, becoming dull green with age. REPRODUCTIVE PARTS Flowers open at same time as leaves in May; tree often covered with white blossom. Fruit up to 1.8cm long, resembling a tiny pear on a long stalk; golden brown at first, ripening later to red and marked by numerous brown lenticels. STATUS AND DISTRIBUTION A scarce native of SW Britain (also found in W France and the Iberian peninsula). Here, it is usually found in hedgerows and copses. COMMENTS Suckering is an important means by which Plymouth Pear reproduces. Unsympathetic hedgerow maintenance, which does not suit the species, helps explain its limited and seemingly diminishing range.

Plymouth Pear fruit

Plymouth Pear leaf

Wild Pear *Pyrus pyraster*

(Rosaceae) 20m

A medium-sized deciduous tree, sometimes becoming fairly large and spreading, but often rounded in outline and no more than 8–10m tall. BARK Dark brown, breaking up into small square plates. BRANCHES May be spreading, or sometimes ascending, and normally spiny; twigs smooth, greyish brown. LEAVES Up to 7cm long, elliptic or rounded, or sometimes heart-shaped near the base, but nearly always with a toothed margin near the apex. Hairless when fully grown; petiole up to 7cm long. REPRODUCTIVE PARTS Long-stalked, white flowers open at same time as leaves in April–May. Petals in clusters of about 5, sometimes looking slightly crushed. The tree is often densely covered with blossom and can be a most attractive sight in woodland in spring. Fruits small, hard, rounded, about 3.5cm in diameter; yellowish brown, sometimes blackened, and pitted with numerous tiny lenticels. They grow on thin stalks. STATUS AND DISTRIBUTION A native of a wide area of Europe, including S Britain. Found in open woodlands and copses. Usually solitary, and typically easy to spot for a short time in spring when the white blossom is open. COMMENTS With many individual trees it can be difficult to ascribe identity or precise parentage with any certainty because Common Pear varieties are occasionally naturalised, and other *Pyrus* species are sometimes planted. Neolithic charcoal made from Wild Pear wood has been found and this perhaps hints at the species' greater abundance in times past. Certainly in medieval times it was important enough to be mentioned in charters and land registers.

Plymouth Pear leaf

Wild Pear fruit

Wild Pear leaf

Plymouth Pear bark

Plymouth Pear foliage

Plymouth Pear fruit

Wild Pear

Wild Pear bark

Wild Pear foliage

Wild Pear flowers

Wild Pear fruits

Willow-leaved
Pear leaf

Willow-leaved
Pear leaf

Willow-leaved Pear *Pyrus salicifolia* (Rosaceae) 10m
Small deciduous tree with a rounded crown. BARK Rough, scaly and usually dark brown. BRANCHES Mostly level with pendulous, very downy twigs. LEAVES Narrow, to 9cm long, like willow leaves: silvery grey on both surfaces at first, but greener on upper surface later in season. REPRODUCTIVE PARTS White flowers, to 2cm across, usually opening with leaves. Fruits about 3cm long, pear-shaped or sometimes more pointed, and brown when ripe, on a downy pedicel. STATUS AND DISTRIBUTION Native of central Asia. Grown in Britain for ornament. COMMENT 'Pendula' is a popular cultivar.

Pyrus elaeagnifolia (Rosaceae) 10m
Small, often slender tree. BARK Rough and scaly. BRANCHES Spreading and spiny; twigs covered with grey hairs. LEAVES Alternate, to 8cm long, lanceolate, sometimes toothed at tip, and covered with thick white down, even at end of growing season. REPRODUCTIVE PARTS White, almost sessile flowers open with leaves. Thick-stalked fruits about 1.3cm long and pear-shaped, sometimes globular, remaining green when ripe. STATUS AND DISTRIBUTION Native of the Balkans eastwards; planted in Britain occasionally.

Pyrus
elaeagnifolia
leaf

Sage-leaved Pear *Pyrus salvifolia* (Rosaceae) 10m
Small, much-branched tree. BARK Rough and scaly. BRANCHES Spreading and spiny with blackish, almost hairless old twigs. LEAVES To 5cm long, elliptical, smooth above, grey and woolly below. REPRODUCTIVE PARTS White flowers open with leaves, followed by pear-shaped fruit, to 8cm long. Pedicel and young fruit are woolly; bitter fruit ripens yellow; used to make perry. STATUS AND DISTRIBUTION Occurs in the wild from France eastwards; planted in Britain occasionally. COMMENT Possibly a hybrid between *P. communis* and *P. nivalis*.

Sage-leaved
Pear leaf

Almond-leaved Pear
Pyrus amygdaliformis (Rosaceae) 6m
Small tree. BARK Rough and scaly. BRANCHES Often dense and sparsely spiny with greyish, woolly young twigs. LEAVES To 8cm long, usually lanceolate with a sparsely toothed margin. Young leaves downy, fully grown leaves shiny above and slightly downy below. REPRODUCTIVE PARTS White flower clusters, opening with the leaves. Thick-stalked fruits rounded, to 3cm across, ripening dark yellow. STATUS AND DISTRIBUTION Native of SE Europe; planted in Britain occasionally.

Almond-leaved
Pear leaf

Snow Pear *Pyrus nivalis* (Rosaceae) 20m
Medium-sized tree. BARK Rough and scaly. BRANCHES Ascending, usually spineless. LEAVES To 9cm long and smooth; blade runs decurrently down petiole. REPRODUCTIVE PARTS White flowers opening just after leaves. Fruits to 5cm long, rounded, greenish yellow with purple dots. STATUS AND DISTRIBUTION Native from France to Russia; planted in Britain occasionally.

Snow Pear leaf

Willow-leaved Pear fruit and foliage

Willow-leaved Pear flowers

Pyrus elaeagnifolia fruit and foliage

Sage-leaved Pear fruit

Sage-leaved Pear bark

Almond-leaved Pear bark

Snow Pear bark

Almond-leaved Pear flowers and foliage

Almond-leaved Pear fruit

Snow Pear foliage

Snow Pear fruit and foliage

APPLES *MALUS* (FAMILY ROSACEAE)

About 25 species occur in northern temperate regions, although there are countless varieties and cultivars used for their highly important edible fruit and sometimes for their attractive blossom. They are hardy trees, growing in a variety of soils and climates, and some produce good-quality timber suitable for turnery. Most important fruiting varieties are propagated by grafting on to healthy stocks.

Cultivated Apple *Malus domestica* (Rosaceae) 15m

Cultivated Apple leaf

A familiar orchard tree producing copious quantities of edible fruits. BARK Usually brown and fissured. BRANCHES Tangled unless pruned. Twigs downy. LEAVES Up to 13cm long, elliptical and rounded at the base with a slightly pointed tip and toothed margin. Slightly downy on upper surface and normally very downy on lower surface. REPRODUCTIVE PARTS Flowers white or tinged with pink and, in some varieties, produced abundantly in short-stalked clusters. Fruits normally larger than 5cm in diameter and indented at the pedicel. A great variety of shapes, sizes, tastes and colours exist. STATUS AND DISTRIBUTION Almost always found in cultivation in orchards and gardens across much of Britain and Ireland. Occasionally naturalised, or found in isolated places where human habitation once occurred, or where apple cores containing seeds ('pips') were discarded. Cultivated Apple is a hybrid species, probably between the Wild Crab *M. sylvestris* and *M. dasyphylla*, and possibly *M. praecox*. COMMENTS There are more than 2,000 varieties of Cultivated Apple that are particular to Britain and Ireland. Popular and traditional varieties include Bramley, Cox's Orange Pippin, Discovery, Russets (many regional forms) and Worcester Pearmain. Timber from the Apple tree is excellent for wood-turning, for making mallet heads, and for imparting a rich fragrance to wood-smoke on log fires. Apple trees in old orchards are often parasitised by Mistletoe *Viscum album*, which becomes very obvious in winter when the leaves have fallen. Apple leaves are the food plant for a range of moth larvae, and those of the Red-belted Clearwing *Synanthedon myopaeformis* (a moth) live and feed in the trunks of old and gnarled trees.

Russet apple

Apple turned from spalted Apple wood

Lord Derby apple

Granny Smith apple

Cultivated Apple

Cultivated Apple blossom

Cultivated Apple fruit

Mistletoe

Discovery apples

Red-belted Clearwing

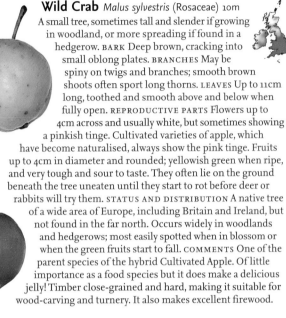

Wild Crab
fruit

Wild Crab
leaf

Malus
'John
Downie'
fruit

Malus
'John
Downie'
leaf

Plum-leaved Crab leaf

Siberian
Crab leaf

Siberian
Crab fruit

Wild Crab *Malus sylvestris* (Rosaceae) 10m

A small tree, sometimes tall and slender if growing in woodland, or more spreading if found in a hedgerow. BARK Deep brown, cracking into small oblong plates. BRANCHES May be spiny on twigs and branches; smooth brown shoots often sport long thorns. LEAVES Up to 11cm long, toothed and smooth above and below when fully open. REPRODUCTIVE PARTS Flowers up to 4cm across and usually white, but sometimes showing a pinkish tinge. Cultivated varieties of apple, which have become naturalised, always show the pink tinge. Fruits up to 4cm in diameter and rounded; yellowish green when ripe, and very tough and sour to taste. They often lie on the ground beneath the tree uneaten until they start to rot before deer or rabbits will try them. STATUS AND DISTRIBUTION A native tree of a wide area of Europe, including Britain and Ireland, but not found in the far north. Occurs widely in woodlands and hedgerows; most easily spotted when in blossom or when the green fruits start to fall. COMMENTS One of the parent species of the hybrid Cultivated Apple. Of little importance as a food species but it does make a delicious jelly! Timber close-grained and hard, making it suitable for wood-carving and turnery. It also makes excellent firewood.

SIMILAR TREE
Malus **'John Downie'** (10m) A popular cultivar, with rather narrow leaves. Fruits maturing bright reddish orange; edible and delicious (unlike Wild Crab), and still good for jelly although without the same degree of sourness.

Plum-leaved Crab *Malus prunifolia* (Rosaceae) 10m

An upright tree, becoming rather straggly when neglected or naturalised. BARK Greyish brown with striking fissures. BRANCHES Dense and irregularly divided. LEAVES Broadly oval and plum-like, to 7cm long, shiny above but rather hairy below when young. REPRODUCTIVE PARTS Flowers up to 5cm across and pinkish, in dense clusters among terminal leaves on shoots. STATUS AND DISTRIBUTION Probably of E Asian origin. Planted rather infrequently in parks and gardens in Britain and Ireland, and very occasionally naturalised.

Siberian Crab *Malus baccata* (Rosaceae) 15m

A domed and rather spreading tree. BARK Brownish and regularly cracking. BRANCHES Much divided and dense. LEAVES Rather slender; matt rather than shiny. REPRODUCTIVE PARTS Compact flower heads of white blossom make this a popular garden tree. Fruit green at first, ripening to bright red and remaining on tree long after leaves have fallen, providing a late feed for winter migrant birds. STATUS AND DISTRIBUTION A native of China, planted in Britain and Ireland in parks and gardens.

Wild Crab bark

Wild Crab flowers

Malus 'John Downie' flowers

Malus 'John Downie' blossom

Wild Crab

Plum-leaved Crab blossom

Siberian Crab fruit and foliage

Plum-leaved Crab fruit

Japanese
Crab leaf

Japanese Crab *Malus floribunda* (Rosaceae) 8m

A compact, densely crowned small tree on a thick bole with dark brown, fissured bark. Twigs slightly pendulous and reddish when young, remaining densely hairy. LEAVES Alternate, up to 8cm long, oval with a pointed tip and a toothed margin. Underside downy when leaves first open, becoming smooth later. REPRODUCTIVE PARTS Flowers fragrant, appearing soon after leaves and usually so dense that they hide the leaves. Buds rich pink at first, becoming paler as they open, and blossom gradually fading to white. Fruits rounded, up to 2.5cm across, but sometimes smaller; ripening yellowish, and often as abundant as the flowers. STATUS AND DISTRIBUTION Probably a hybrid between 2 Japanese garden species, as this tree has not been found in the wild. Frequently planted in gardens and parks all over Europe for its attractive blossom and convenient small size.

Japanese
Crab fruit

Hubei
Crab leaf

Hubei Crab *Malus hupehensis* (Rosaceae) 15m

Broadly domed and spreading tree. BARK Reddish brown with scaly plates. BRANCHES Long and spreading, lower ones with shoots that almost reach the ground. LEAVES Narrowly ovate, pointed at the tip, to 10cm long, shiny green above, on a rather long, downy petiole. REPRODUCTIVE PARTS Flowers white, buds pink. Fruits reddish, 1cm across. STATUS AND DISTRIBUTION Scarce native of E Asia, widely planted in British parks and gardens.

Hubei Crab bark

Hawthorn-leaved Crab *Malus florentina* (Rosaceae) 10m

Attractive and rather conical small tree with good blossom, small fruits and colourful autumn foliage. BARK Brown with yellowish scales. BRANCHES Mostly level to upright. LEAVES To 8cm long, sharply lobed, resembling those of Wild Service-tree. REPRODUCTIVE PARTS Flowers whitish and fruits red, around 1cm across. STATUS AND DISTRIBUTION Possibly a hybrid between a *Malus* species and Wild Service-tree *Sorbus torminalis,* which it resembles (*see* p. 220). Planted occasionally.

Hawthorn-leaved
Crab fruit

Hawthorn-leaved
Crab leaf

Cherry-crab Hybrid *Malus × zumi* (Rosaceae) 9m

Hybrid crab apple popular for its small red fruits. BARK Reddish brown. BRANCHES Spreading. LEAVES Narrowly oval. REPRODUCTIVE PARTS Flowers white, buds pink. Fruits ovoid, to 2cm long; typically red but lemon-yellow in cultivar 'Golden Hornet'. STATUS AND DISTRIBUTION Widely planted.

SIMILAR TREES

Cherry-crab *Malus × robusta* (10m) Buds red, flowers whitish and fruits reddish and cherry-like.

Purple Crab *Malus × purpurea* (10m) Leaves tinged purple, flowers reddish and fruits deep red.

Weeping Purple Crab *Malus × gloriosa* (10m) Similar to Purple Crab but with weeping foliage.

Japanese Crab fruit and foliage

Japanese Crab flowers

Hubei Crab flowers

Hubei Crab ripening fruits

Hawthorn-leaved Crab fruits and foliage

Hawthorn-leaved Crab

Cherry-crab hybrid fruits

WHITEBEAMS AND ROWANS SORBUS (FAMILY ROSACEAE)

About 80 species of medium to small trees and shrubs, mostly with showy umbels of white flowers and clusters of colourful berries. Rowans typically have pinnate leaves and whitebeams have simple, usually toothed leaves. They hybridise and confusing leaf forms occur. Some very rare and local *Sorbus* species exist, native to Britain and Ireland and restricted to isolated localities. These attract considerable attention from conservationists.

Rowan (Mountain Ash) *Sorbus aucuparia* (Rosaceae) 20m

Rowan leaf

A small to medium-sized deciduous tree with a fairly open, domed crown. BARK Silvery grey; usually smooth but sometimes feels slightly ridged. BRANCHES Ascending and quite widely spaced with purple-tinged twigs that are hairy when young but become smooth later. Buds oval with curved tips; mostly purple and covered with greyish hairs. LEAVES Compound and pinnate, composed of 5–8 pairs of toothed leaflets, each one up to 6cm long, ovoid and markedly toothed. The central rachis is rounded near the base, and grooved between the leaflets. REPRODUCTIVE PARTS Flowers in dense heads in May; each flower is up to 1cm in diameter with 5 creamy white petals. Fruits rounded, under 1cm long and bright scarlet, hanging in colourful clusters and persisting after the leaves have fallen. STATUS AND DISTRIBUTION A

Rowan fruit

native of a wide area of Europe, including Britain and Ireland. It occurs in woodland and open land on a variety of soils, apart from very wet ones, and grows at a higher altitude on mountains than many other species. Often planted as a town tree in squares and along roadsides. COMMENTS The bright scarlet fruits are a favourite food of birds in winter and often attract migrants like Waxwings into busy town centres.

Service-tree *Sorbus domestica* (Rosaceae) 20m

Resembles Rowan but note subtle differences in bark, buds and fruit. BARK Rich brown, fissured, ridged, and often peeling in vertical shreds.

Service-tree fruit

BRANCHES Upright to spreading. Buds smooth, rounded and green, unlike the purple, pointed buds of Rowan.

Service-tree leaf

LEAVES Alternate and pinnate, composed of up to 8 pairs of oblong leaflets about 5cm long, toothed only on outer half (almost all round margin in Rowan) and softly hairy on underside. REPRODUCTIVE PARTS Flowers in May, in rounded, branched clusters; each flower about 1.5cm across and composed of 5 creamy white petals. Small pear- or sometimes apple-shaped fruits, up to 2cm long and green or brown like a russet apple. They have a very sharp taste when ripe, but after a frost they become more palatable. STATUS AND DISTRIBUTION Widespread in S Europe but a rare British native, mainly in Bristol Channel area; also planted occasionally.

Service-tree fruit

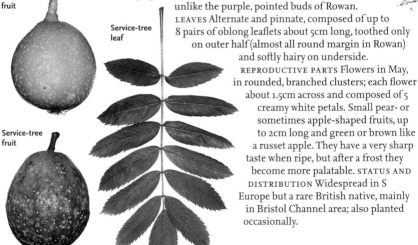

Rowan fruits

Rowan tree with fruits

Rowan flowers and foliage

Service-tree foliage and faded flowers

Rowan bark

Service-tree bark

Service-tree fruits and foliage

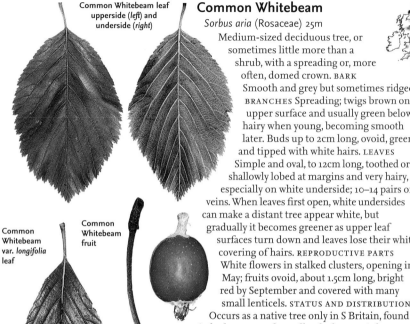

Common Whitebeam leaf upperside (*left*) and underside (*right*)

Common Whitebeam var. *longifolia* leaf

Common Whitebeam fruit

Common Whitebeam

Sorbus aria (Rosaceae) 25m

Medium-sized deciduous tree, or sometimes little more than a shrub, with a spreading or, more often, domed crown. BARK Smooth and grey but sometimes ridged. BRANCHES Spreading; twigs brown on upper surface and usually green below; hairy when young, becoming smooth later. Buds up to 2cm long, ovoid, green and tipped with white hairs. LEAVES Simple and oval, to 12cm long, toothed or shallowly lobed at margins and very hairy, especially on white underside; 10–14 pairs of veins. When leaves first open, white undersides can make a distant tree appear white, but gradually it becomes greener as upper leaf surfaces turn down and leaves lose their white covering of hairs. REPRODUCTIVE PARTS White flowers in stalked clusters, opening in May; fruits ovoid, about 1.5cm long, bright red by September and covered with many small lenticels. STATUS AND DISTRIBUTION Occurs as a native tree only in S Britain, found in hedgerows and woodland edges, mainly on limestone and other calcareous soils; often found on chalk downland slopes. However, its true native range is confused because it is often planted in towns and along roadside verges, being relatively tolerant of atmospheric pollution, and is naturalised in the countryside at large. COMMENTS A number of cultivars are commonly planted in gardens and streets. 'Lutescens' has purple twigs, smaller leaves and a very neat, compact habit; fruits are woolly before they ripen. 'Decaisneana' has larger leaves, dark and glossy above, and fruits are white-spotted on yellow stalks. 'Majestica' has larger, thicker leaves, sometimes up to 15cm long, and stronger, more upright branches. Common Whitebeam fruits are popular with birds in autumn and winter. Fieldfares and Redwings find them irresistible in the countryside, and urban-planted cultivars lure Waxwings into towns. The timber is hard and durable: in the past, before cast iron was widely available, it was used to make machinery cogs for mills and the like. It also makes durable tool handles and is used in turnery. The species' English name is derived from the tree's white appearance in spring, and from *beam*, the Saxon word for tree.

SIMILAR TREES

Bastard Service *Sorbus* × *thuringiaca* (16m) A naturally occurring hybrid between Common Whitebeam and Rowan. Leaves with 2–3 pairs of free lobes; fruits brown. *S. thibetica* has leaves to 20cm long and fruits to 2cm long; a Himalayan species, planted in Britain for ornament.

NOTE As with other *Sorbus* species, the classification of *Sorbus aria* is a minefield, in a state of flux, and viewed differently by various experts. Some treat *Sorbus aria* as an aggregate species that includes some of the rare and regionally restricted species described on the following pages. Here, it is treated as a distinct species in its own right.

Common Whitebeam bark

Common Whitebeam fruits

Common Whitebeam

Common Whitebeam flowers and foliage

Common Whitebeam leaves underside

stard Service
f

Bastard Service fruit

Sorbus thibetica leaf

Sorbus thibetica fruit

Wild Service-tree *Sorbus torminalis* (Rosaceae) 25m

A medium-sized deciduous tree with a spreading or sometimes domed habit, depending on its situation. BARK Finely fissured into squarish brown plates, thought to be the origin of one of its vernacular English names: 'Chequers'. BRANCHES Straight and spreading with twigs that are shiny and usually dark brown, terminating in rounded shiny green buds about 5mm long. LEAVES Up to 10cm long with 3–5 pairs of pointed lobes and a sharply toothed margin; the basal lobes projecting at right angles, the other lobes pointed forwards. The leaves are amongst the first to change colour in the autumn, producing a rich display of reds and russets, and making the tree easy to find at this season. REPRODUCTIVE PARTS White flowers, in May, up to 1.5cm in diameter, in loose clusters on woolly pedicels. Fruits up to 1.8cm in diameter, rounded and brownish, with numerous lenticels in the skin, and normally ripe in September. STATUS AND DISTRIBUTION A native of much of Europe, including Britain, and normally found in copses and woodlands, especially ancient woodlands on heavier soils. The tree is now rather scarce, partly because of unsympathetic woodland management and because most British seeds stubbornly refuse to germinate (they need hot summer temperatures to do so). Consequently, Wild Service-tree is often used as an indicator of relatively undisturbed ancient semi-natural woodland. The species is also planted occasionally in parks and gardens, mainly for its glorious, if brief and rather unreliable, autumn colours. It is also used in woodland replanting schemes that involve native species. COMMENTS There are various explanations for the origin of the word 'service', used in this species' name and that of Service-tree *Sorbus domestica*. The most plausible is that it is derived from *cerevisia*, the Latin name for beer – in the past, Wild Service-tree fruits were used to sweeten beer, and used on their own to produce a potent alcoholic beverage called 'checkers'; this links with the species' vernacular name and probably explains why so many rural pubs are so named. Quite what this beverage would have tasted like, or what its effect would have been, is open to speculation since early physicians used infusions made from the fruits to treat colic and dysentery. The timber from Wild Service-tree is hard but the scarcity of the tree means it has no specific uses these days. In the past, however, it was used to make charcoal.

Wild Service-tree leaf

Wild Service-tree fruit

Sorbus × vagensis fruit

Sorbus × vagensis leaf

Sorbus × vagensis is a naturally occurring hybrid between *S. aria* and *S. torminalis*

Wild Service-tree bark

Wild Service-tree flowers and foliage

Wild Service-tree growing in hedgerow

Wild Service-tree fruits

Sorbus × vagensis foliage

NOTE Some authors treat regional *Sorbus* species (pp. 222–9) as 'microspecies', lumping them within the following species aggregates: Broad-leaved Whitebeam *Sorbus latifolia* agg., Swedish Whitebeam *Sorbus intermedia* agg. or Common Whitebeam *Sorbus aria* agg.

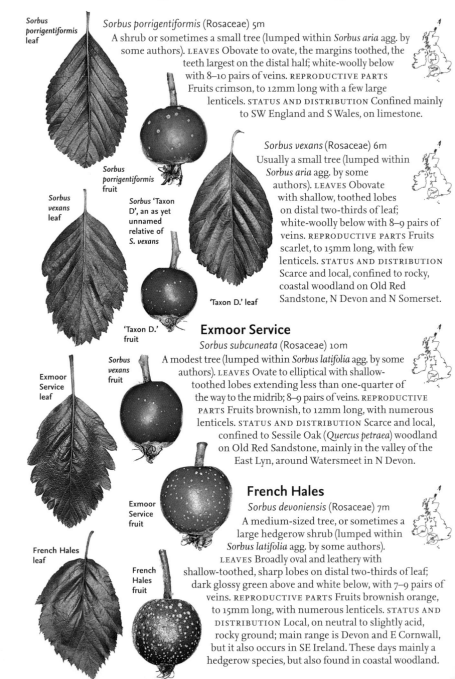

Sorbus porrigentiformis leaf

Sorbus porrigentiformis (Rosaceae) 5m
A shrub or sometimes a small tree (lumped within *Sorbus aria* agg. by some authors). LEAVES Obovate to ovate, the margins toothed, the teeth largest on the distal half; white-woolly below with 8–10 pairs of veins. REPRODUCTIVE PARTS Fruits crimson, to 12mm long with a few large lenticels. STATUS AND DISTRIBUTION Confined mainly to SW England and S Wales, on limestone.

Sorbus porrigentiformis fruit

Sorbus vexans leaf

Sorbus 'Taxon D', an as yet unnamed relative of *S. vexans*

'Taxon D.' leaf

Sorbus vexans (Rosaceae) 6m
Usually a small tree (lumped within *Sorbus aria* agg. by some authors). LEAVES Obovate with shallow, toothed lobes on distal two-thirds of leaf; white-woolly below with 8–9 pairs of veins. REPRODUCTIVE PARTS Fruits scarlet, to 15mm long, with few lenticels. STATUS AND DISTRIBUTION Scarce and local, confined to rocky, coastal woodland on Old Red Sandstone, N Devon and N Somerset.

'Taxon D.' fruit

Sorbus vexans fruit

Exmoor Service

Sorbus subcuneata (Rosaceae) 10m
A modest tree (lumped within *Sorbus latifolia* agg. by some authors). LEAVES Ovate to elliptical with shallow-toothed lobes extending less than one-quarter of the way to the midrib; 8–9 pairs of veins. REPRODUCTIVE PARTS Fruits brownish, to 12mm long, with numerous lenticels. STATUS AND DISTRIBUTION Scarce and local, confined to Sessile Oak (*Quercus petraea*) woodland on Old Red Sandstone, mainly in the valley of the East Lyn, around Watersmeet in N Devon.

Exmoor Service leaf

Exmoor Service fruit

French Hales

Sorbus devoniensis (Rosaceae) 7m
A medium-sized tree, or sometimes a large hedgerow shrub (lumped within *Sorbus latifolia* agg. by some authors). LEAVES Broadly oval and leathery with shallow-toothed, sharp lobes on distal two-thirds of leaf; dark glossy green above and white below, with 7–9 pairs of veins. REPRODUCTIVE PARTS Fruits brownish orange, to 15mm long, with numerous lenticels. STATUS AND DISTRIBUTION Local, on neutral to slightly acid, rocky ground; main range is Devon and E Cornwall, but it also occurs in SE Ireland. These days mainly a hedgerow species, but also found in coastal woodland.

French Hales leaf

French Hales fruit

Sorbus porrigentiformis flowers and foliage

Sorbus vexans flowers and foliage

Exmoor Service flowers and foliage

French Hales bark

French Hales flowers and foliage

French Hales fruit and foliage

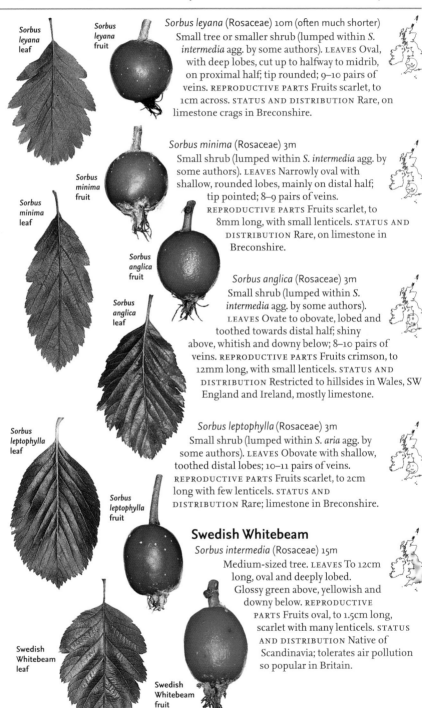

Sorbus leyana leaf

Sorbus leyana fruit

Sorbus leyana (Rosaceae) 10m (often much shorter) Small tree or smaller shrub (lumped within *S. intermedia* agg. by some authors). LEAVES Oval, with deep lobes, cut up to halfway to midrib, on proximal half; tip rounded; 9–10 pairs of veins. REPRODUCTIVE PARTS Fruits scarlet, to 1cm across. STATUS AND DISTRIBUTION Rare, on limestone crags in Breconshire.

Sorbus minima fruit

Sorbus minima leaf

Sorbus minima (Rosaceae) 3m Small shrub (lumped within *S. intermedia* agg. by some authors). LEAVES Narrowly oval with shallow, rounded lobes, mainly on distal half; tip pointed; 8–9 pairs of veins. REPRODUCTIVE PARTS Fruits scarlet, to 8mm long, with small lenticels. STATUS AND DISTRIBUTION Rare, on limestone in Breconshire.

Sorbus anglica fruit

Sorbus anglica leaf

Sorbus anglica (Rosaceae) 3m Small shrub (lumped within *S. intermedia* agg. by some authors). LEAVES Ovate to obovate, lobed and toothed towards distal half; shiny above, whitish and downy below; 8–10 pairs of veins. REPRODUCTIVE PARTS Fruits crimson, to 12mm long, with small lenticels. STATUS AND DISTRIBUTION Restricted to hillsides in Wales, SW England and Ireland, mostly limestone.

Sorbus leptophylla leaf

Sorbus leptophylla fruit

Sorbus leptophylla (Rosaceae) 3m Small shrub (lumped within *S. aria* agg. by some authors). LEAVES Obovate with shallow, toothed distal lobes; 10–11 pairs of veins. REPRODUCTIVE PARTS Fruits scarlet, to 2cm long with few lenticels. STATUS AND DISTRIBUTION Rare; limestone in Breconshire.

Swedish Whitebeam

Sorbus intermedia (Rosaceae) 15m Medium-sized tree. LEAVES To 12cm long, oval and deeply lobed. Glossy green above, yellowish and downy below. REPRODUCTIVE PARTS Fruits oval, to 1.5cm long, scarlet with many lenticels. STATUS AND DISTRIBUTION Native of Scandinavia; tolerates air pollution so popular in Britain.

Swedish Whitebeam leaf

Swedish Whitebeam fruit

Sorbus leyana foliage

Sorbus minima flowers and foliage

Sorbus anglica flowers and foliage

Sorbus leptophylla flowers and foliage

Swedish Whitebeam

Swedish Whitebeam flowers and foliage

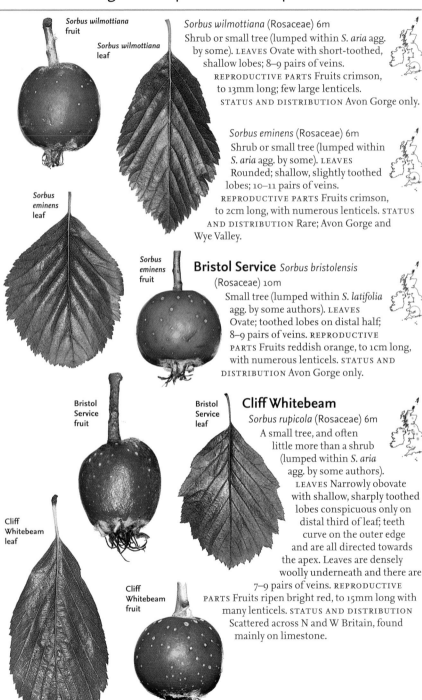

Sorbus wilmottiana fruit

Sorbus wilmottiana leaf

Sorbus wilmottiana (Rosaceae) 6m
Shrub or small tree (lumped within *S. aria* agg. by some). LEAVES Ovate with short-toothed, shallow lobes; 8–9 pairs of veins.
REPRODUCTIVE PARTS Fruits crimson, to 13mm long; few large lenticels.
STATUS AND DISTRIBUTION Avon Gorge only.

Sorbus eminens (Rosaceae) 6m
Shrub or small tree (lumped within *S. aria* agg. by some). LEAVES Rounded; shallow, slightly toothed lobes; 10–11 pairs of veins.
REPRODUCTIVE PARTS Fruits crimson, to 2cm long, with numerous lenticels. STATUS AND DISTRIBUTION Rare; Avon Gorge and Wye Valley.

Sorbus eminens leaf

Sorbus eminens fruit

Bristol Service *Sorbus bristolensis*
(Rosaceae) 10m
Small tree (lumped within *S. latifolia* agg. by some authors). LEAVES Ovate; toothed lobes on distal half; 8–9 pairs of veins. REPRODUCTIVE PARTS Fruits reddish orange, to 1cm long, with numerous lenticels. STATUS AND DISTRIBUTION Avon Gorge only.

Bristol Service fruit

Bristol Service leaf

Cliff Whitebeam
Sorbus rupicola (Rosaceae) 6m
A small tree, and often little more than a shrub (lumped within *S. aria* agg. by some authors).
LEAVES Narrowly obovate with shallow, sharply toothed lobes conspicuous only on distal third of leaf; teeth curve on the outer edge and are all directed towards the apex. Leaves are densely woolly underneath and there are 7–9 pairs of veins. REPRODUCTIVE PARTS Fruits ripen bright red, to 15mm long with many lenticels. STATUS AND DISTRIBUTION Scattered across N and W Britain, found mainly on limestone.

Cliff Whitebeam leaf

Cliff Whitebeam fruit

Sorbus wilmottiana flowers and foliage

Bristol Service flowers and foliage

Sorbus eminens flowers and foliage

Bristol Service bark

Cliff Whitebeam flowers and foliage

Bristol Service fruits

Cliff Whitebeam fruits

Arran
Service-tree
leaf

Arran
Service-tree
fruit

Arran Service-tree

Sorbus pseudofennica (Rosaceae) 7m

An upright tree (treated as a 'microspecies' of Bastard Service-tree *S. × thuringiaca* in some books). LEAVES Pinnately divided with 7–9 pairs of veins; proximal 1–2 pairs of leaflets are free. REPRODUCTIVE PARTS Fruits scarlet, to 12mm long, with few lenticels. STATUS AND DISTRIBUTION More or less restricted to a single site on the Scottish island of Arran, growing on granite, with a population of around just 400 plants.

Arran
Whitebeam
fruit

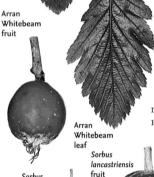

Arran Whitebeam

Sorbus arranensis (Rosaceae) 7m

A rather domed tree (a stable and fertile hybrid between Rowan *S. aucuparia* and Cliff Whitebeam *S. rupicola*, lumped within *S. intermedia* agg. by some authors). LEAVES Narrowly oval to elliptical in overall outline but with finely toothed rounded lobes, cut roughly halfway to midrib; 7–8 pairs of veins. REPRODUCTIVE PARTS Fruits red, to 10mm long, with few lenticels. STATUS AND DISTRIBUTION Restricted to granite stream sides on Arran, the population numbering around 400.

Arran
Whitebeam
leaf

Sorbus lancastriensis fruit

Sorbus lancastriensis leaf

Sorbus lancastriensis (Rosaceae) 5m

A shrub or small tree (lumped within *S. aria* agg. by some authors). LEAVES Ovate, the margin with short, sharp teeth; 8–10 pairs of veins. Downy below. REPRODUCTIVE PARTS Flowers and fruits on slightly downy stalks. Fruits red, to 1.5cm long, with prominent lenticels when ripe. STATUS AND DISTRIBUTION Mainly confined to limestone rocks in NW England.

Service-tree of
Fontainebleau
fruit

Service-tree of
Fontainebleau
leaf

Service-tree of Fontainebleau

Sorbus latifolia (Rosaceae) 18m

Spreading, divided tree. LEAVES To 10cm long, broadly oval with indistinct triangular lobes and double-toothed margins. Shiny green above, downy grey below on 7–9 pairs of veins. REPRODUCTIVE PARTS Fruits rounded, to 1.5cm long, yellowish brown with large lenticels. STATUS AND DISTRIBUTION Probably a hybrid between Whitebeam and Wild Service-tree, first found near Fontainebleau, France. Widely planted; also naturalised.

Sorbus hibernica fruit

Sorbus hibernica leaf

Sorbus hibernica (Rosaceae) 6m

A shrub or small tree (lumped within *S. aria* agg. by some authors). LEAVES Ovate to slightly obovate, with toothed lobes on distal half of leaf; whitish woolly below, and with 9–11 pairs of veins. REPRODUCTIVE PARTS Fruits pinkish red, to 15mm long, with few lenticels. STATUS AND DISTRIBUTION Confined to areas of scrub and woodland on limestone in central Ireland.

Arran Service-tree flowers and foliage

Arran Service-tree fruit and autumn leaves

Arran Whitebeam flowers and foliage

Sorbus lancastriensis flowers and foliage

Sorbus hibernica flowers and foliage

Service-tree of Fontainebleau flowers and foliage

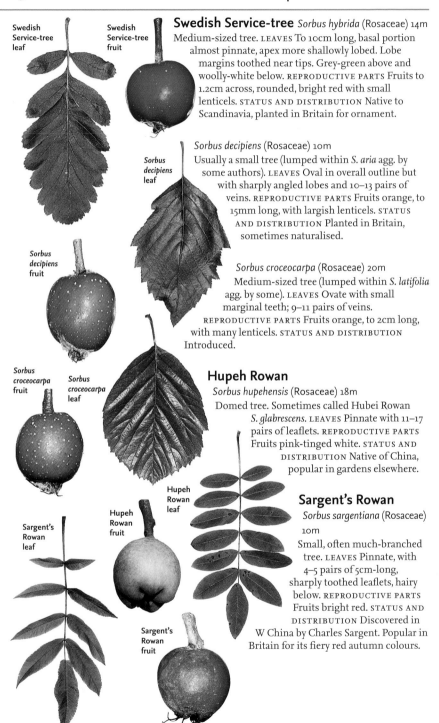

Swedish
Service-tree
leaf

Swedish
Service-tree
fruit

Swedish Service-tree *Sorbus hybrida* (Rosaceae) 14m

Medium-sized tree. LEAVES To 10cm long, basal portion almost pinnate, apex more shallowly lobed. Lobe margins toothed near tips. Grey-green above and woolly-white below. REPRODUCTIVE PARTS Fruits to 1.2cm across, rounded, bright red with small lenticels. STATUS AND DISTRIBUTION Native to Scandinavia, planted in Britain for ornament.

*Sorbus
decipiens
leaf*

Sorbus decipiens (Rosaceae) 10m

Usually a small tree (lumped within *S. aria* agg. by some authors). LEAVES Oval in overall outline but with sharply angled lobes and 10–13 pairs of veins. REPRODUCTIVE PARTS Fruits orange, to 15mm long, with largish lenticels. STATUS AND DISTRIBUTION Planted in Britain, sometimes naturalised.

*Sorbus
decipiens
fruit*

Sorbus croceocarpa (Rosaceae) 20m

Medium-sized tree (lumped within *S. latifolia* agg. by some). LEAVES Ovate with small marginal teeth; 9–11 pairs of veins. REPRODUCTIVE PARTS Fruits orange, to 2cm long, with many lenticels. STATUS AND DISTRIBUTION Introduced.

Sorbus
croceocarpa
fruit

*Sorbus
croceocarpa
leaf*

Hupeh Rowan

Sorbus hupehensis (Rosaceae) 18m

Domed tree. Sometimes called Hubei Rowan *S. glabrescens*. LEAVES Pinnate with 11–17 pairs of leaflets. REPRODUCTIVE PARTS Fruits pink-tinged white. STATUS AND DISTRIBUTION Native of China, popular in gardens elsewhere.

Hupeh
Rowan
leaf

Hupeh
Rowan
fruit

Sargent's Rowan

Sorbus sargentiana (Rosaceae) 10m

Small, often much-branched tree. LEAVES Pinnate, with 4–5 pairs of 5cm-long, sharply toothed leaflets, hairy below. REPRODUCTIVE PARTS Fruits bright red. STATUS AND DISTRIBUTION Discovered in W China by Charles Sargent. Popular in Britain for its fiery red autumn colours.

Sargent's
Rowan
leaf

Sargent's
Rowan
fruit

Swedish Service-tree fruits and foliage

Sorbus decipiens flowers and foliage

Sorbus croceocarpa flowers and foliage

Hupeh Rowan fruits and foliage

Sargent's Rowan flowers and foliage

Loquat *Eriobotrya japonica* (Rosaceae) 10m

Loquat
leaf

Small evergreen tree or large shrub. BARK Grey-buff. BRANCHES Thick, with hairy twigs. LEAVES To 25cm long, elliptical, toothed with distinct veins; leathery, glossy green above, downy reddish brown below. REPRODUCTIVE PARTS Flowers white, to 1cm across, in branched, downy terminal spikes. Fruits rounded, yellow, fleshy, to 6cm long. STATUS AND DISTRIBUTION Native of China, introduced to Britain for ornament but not hardy.

SNOWY MESPILS *AMELANCHIER* (FAMILY ROSACEAE)

Small trees or large shrubs with spineless branches and alternate leaves. Flowers usually in lax terminal spikes, and normally white. Fruits rounded, usually juicy, sweet and edible; retain dried calyx at tip.

Snowy Mespil
autumn leaf

Snowy Mespil *Amelanchier ovalis* (Rosaceae) 5m

Small deciduous tree or shrub. LEAVES To 5cm long with coarsely toothed margins and downy undersides when first open. REPRODUCTIVE PARTS Flowers in upright spikes of up to 8 white-petalled flowers; fruits blue-black. STATUS AND DISTRIBUTION Native of mainland Europe eastwards; planted in Britain occasionally.

American
Snowy
Mespil
leaf

SIMILAR TREES

American Snowy Mespil *A. laevis* (20m) Small, smooth-barked tree with similar leaves to Snowy Mespil, turning bright red in autumn. Flowers in drooping spikes; fruits 6mm long, purple when ripe. Naturalised in parts of S England, but usually seen in gardens.
Canadian Snowy Mespil *A. canadensis* (15m) Leaves mostly oblong. Native of NE North America, planted in British gardens.

Canadian
Snowy
Mespil
leaf

Juneberry
leaf

Juneberry *Amelanchier lamarckii* (Rosaceae) 9m

Small deciduous, spreading tree with hairy young twigs. LEAVES To 7cm long, elliptical and finely toothed. Tinged purple and woolly at first, smooth and green when older. REPRODUCTIVE PARTS Flowers in drooping, slightly hairy spikes; petals white, to 1.8cm long. Fruits deep purple when ripe, with dried sepals at tip. STATUS AND DISTRIBUTION Probably a hybrid between 2 similar wild species, first arising in North America. Planted in Britain and sometimes naturalised.

Tree
Cotoneaster
leaf

COTONEASTERS *COTONEASTER* (FAMILY ROSACEAE)

About 70 species of shrubs and a single tree. Leaves are normally entire and twigs are thornless. Flowers are usually white and fruit is a berry.

Tree Cotoneaster *Cotoneaster frigidus* (Rosaceae) 20m

Medium-sized tree or many-stemmed shrub. BARK Pale grey. BRANCHES Spineless. LEAVES Elliptical with entire margins, dark green above, hairy and white below. Leaves persist and are semi-evergreen. REPRODUCTIVE PARTS Flowers small, but grouped into dense white clusters about 5cm across. Fruits, to 5mm long, round and red. STATUS AND DISTRIBUTION Native of Himalayas, planted here, sometimes naturalised. Other ornamental species are widely grown, notably *C. watereri*.

Loquat foliage

Loquat fruits and foliage

Snowy Mespil flowers

Snowy Mespil fruits and foliage

Canadian Snowy Mespil fruits and foliage

Juneberry foliage

Juneberry autumn colours

Tree Cotoneaster bark

Tree Cotoneaster fruits and foliage

Tree Cotoneaster flowers and foliage

THORNS CRATAEGUS (FAMILY ROSACEAE)

A large and rather confusing group with many very similar species and numerous local forms. They are characterised by thorny twigs, and many also have attractive flowers and fruits, and good autumn colours. They are usually tough and resilient and survive in harsh conditions, including city centres. Most are large shrubs or small trees, and can be pruned and trained into good stockproof hedges.

Common Hawthorn *Crataegus monogyna* (Rosaceae) 15m

A small, spreading deciduous tree or hedgerow shrub. It may grow on a single stout bole, or be multi-stemmed with a spreading crown. BARK Usually heavily fissured into a fairly regular pattern of vertical grooves; outer layers greyish brown, lower layers more orange. BRANCHES Usually densely packed, as are twigs. Numerous sharp spines. LEAVES To 4.5cm long, roughly ovate and deeply lobed, usually with 3 segments. Lobes pointed, with just a few teeth near the apex. Leaves have a tough feel and are dark green above, paler below, with a few tufts of hairs at the axils of the veins. Petiole about 2cm long and tinged pink. REPRODUCTIVE PARTS Flowers in late spring, in flat-topped clusters of about 10–18 white or sometimes pink-tinged flowers. Hawthorns growing in the open, and receiving plenty of sunlight, often flower prolifically and produce a heavy scent. Each flower is about 1.5cm in diameter with a single style in the centre. The fruits ('haws') are usually rounded, sometimes more ovate, bright red or sometimes darkening to maroon; they contain a hard-cased seed. STATUS AND DISTRIBUTION A very common native across most of Europe, including Britain and Ireland. Common Hawthorn is found in woods and hedgerows, and rapidly colonises waste ground, roadsides and other disturbed habitats, the seeds being bird-sown; it is particularly abundant in drier limestone habitats, and on chalk slopes. COMMENTS Common Hawthorn hybridises readily with other *Crataegus* species and identification can be difficult in regions where they occur together. It is the food plant for the larvae of a wide range of moth species, notably those of the Lappet Moth; leaves and berries are food for the Hawthorn Shield Bug *Acanthosoma haemarrhoidale*. The berries are an important winter food for birds such as Fieldfare and Redwing, and for small mammals including Bank Vole and Wood Mouse. Common Hawthorn is a frequent component of hedgerows and can be used to create a stockproof barrier. In many locations, its abundance can be traced back to widespread hedgerow planting schemes that took place as a consequence of the enclosure movement of land, mostly between the 16th and 18th centuries.

Common Hawthorn leaf

Common Hawthorn fruits

Hawthorn Shield Bug

Common Hawthorn bark

Common Hawthorn flowers

Common Hawthorn fruits

Common Hawthorn

Common Hawthorn berry-covered bush

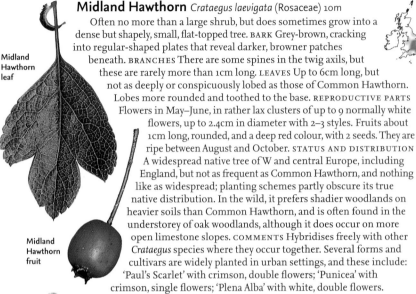

Midland Hawthorn leaf

Midland Hawthorn fruit

Midland Hawthorn *Crataegus laevigata* (Rosaceae) 10m

Often no more than a large shrub, but does sometimes grow into a dense but shapely, small, flat-topped tree. BARK Grey-brown, cracking into regular-shaped plates that reveal darker, browner patches beneath. BRANCHES There are some spines in the twig axils, but these are rarely more than 1cm long. LEAVES Up to 6cm long, but not as deeply or conspicuously lobed as those of Common Hawthorn. Lobes more rounded and toothed to the base. REPRODUCTIVE PARTS Flowers in May–June, in rather lax clusters of up to 9 normally white flowers, up to 2.4cm in diameter with 2–3 styles. Fruits about 1cm long, rounded, and a deep red colour, with 2 seeds. They are ripe between August and October. STATUS AND DISTRIBUTION A widespread native tree of W and central Europe, including England, but not as frequent as Common Hawthorn, and nothing like as widespread; planting schemes partly obscure its true native distribution. In the wild, it prefers shadier woodlands on heavier soils than Common Hawthorn, and is often found in the understorey of oak woodlands, although it does occur on more open limestone slopes. COMMENTS Hybridises freely with other *Crataegus* species where they occur together. Several forms and cultivars are widely planted in urban settings, and these include: 'Paul's Scarlet' with crimson, double flowers; 'Punicea' with crimson, single flowers; 'Plena Alba' with white, double flowers.

Oriental Hawthorn *Crataegus laciniata* (Rosaceae) 10m

Oriental Hawthorn fruit

Rather low and spreading tree. BARK Scaly, brown with pinkish patches. BRANCHES Often twisted; young twigs and pedicels covered with white hairs; becoming smooth and blackish with age. LEAVES Deeply lobed, to 4cm long, with fine white hairs on both sides. REPRODUCTIVE PARTS Flowers creamy white, in dense clusters of up to 16; fruits hairy at first, ripening to orange or red, and containing 3–5 seeds. STATUS AND DISTRIBUTION Native to SE Europe, Spain and Sicily, and occasionally planted in Britain for ornament.

Oriental Hawthorn leaf

SIMILAR TREE

Various-leaved Hawthorn *C. heterophylla* (10m) Some leaves like Oriental Hawthorn, others unlobed. Introduced to parks and gardens.

Azarole leaf

Azarole fruit

Azarole *Crataegus azarolus* (Rosaceae) 8m

A low, spreading tree. BARK Scaly and brown. BRANCHES Twisted and spreading, with numerous thorns. LEAVES As deeply lobed as Oriental Hawthorn but with broader lobes; fine white hairs on both leaf surfaces. REPRODUCTIVE PARTS Flowers creamy white, in dense clusters; fruits yellow, to 2cm across. STATUS AND DISTRIBUTION A native of the Mediterranean region and planted in Britain occasionally for ornament.

Midland Hawthorn flowers

Midland Hawthorn blossom

Midland Hawthorn fruits

Midland Hawthorn 'Rosea Flora Plena'

Oriental Hawthorn flowers and foliage

Oriental Hawthorn fruits and foliage

Azarole flowers and foliage

Azarole fruits

Cockspurthorn *Crataegus crus-galli* (Rosaceae) 10m

Cockspurthorn leaf

A small, usually spreading deciduous tree with a flattish crown and a short bole. BARK Smooth and greyish brown in young trees; fissured in older trees. BRANCHES Purple-brown twigs with numerous sharp spines 7–10cm long. LEAVES Up to 8cm long and about 3cm wide, increasing in width beyond the middle, with a toothed margin. Both surfaces smooth and shiny, dark green in summer and turning rich orange in autumn, often before other species have started to show colour changes. REPRODUCTIVE PARTS White flowers about 1.5cm in diameter, in loose clusters, opening in May. Red globular fruits, ripe in October, persisting after leaves have fallen. STATUS AND DISTRIBUTION A native tree in NE North America, and often planted in Britain and Ireland as a garden or roadside tree, mostly for its striking orange autumn colours. However, not as commonly planted as the 2 trees that follow.

Broad-leaved Cockspurthorn, autumn leaves

Broad-leaved Cockspurthorn

Crataegus persimilis (Rosaceae) 8m

A small, spreading tree with a short bole and a rounded crown. BARK Scaly brown, sometimes with spiral ridges. BRANCHES Twigs and 2cm-long, stiff thorns deep, glossy purple-brown. LEAVES Oval and toothed, smooth and shiny; turning from glossy green, through yellow, orange and copper to deep red by the end of autumn before they finally fall. REPRODUCTIVE PARTS Flowers white, in clusters; fruits rounded and red, up to 1.5cm long. STATUS AND DISTRIBUTION A North American species, popularly planted in Britain on roadsides and in gardens (typically as the cultivar 'Prunifolia'); it also makes a good stockproof hedge.

Pear-fruited Cockspurthorn leaf

Pear-fruited Cockspurthorn fruit

Hybrid Cockspurthorn

Crataegus × lavalleei (Rosaceae) 12m

A dense, domed to spreading tree. BARK Grey and scaly. BRANCHES Level, with twigs growing thickly on the upper side, a feature that makes winter recognition easy. LEAVES Narrow and glossy green, turning dark red late in the autumn. REPRODUCTIVE PARTS Flowers white, in clusters; fruits dull red, to 18mm long. STATUS AND DISTRIBUTION A common tree in town gardens and on roadsides.

SIMILAR TREES
Pear-fruited Cockspurthorn *C. pedicellata* (7m) Leaves broadly ovate, lobed and double-toothed; fruits red, pear-shaped, to 2cm long; spines 3–5cm long.
Hairy Cockspurthorn *C. submollis* (8m) Leaves ovate, lobed and double-toothed; fruits red, pear-shaped, to 2cm long; spines 5–7cm long.
Scarlet Thorn *C. mollis* (10m) Leaves broadly triangular with distinct, toothed lobes; fruits reddish. Several other related *Crataegus* species occur in parks and gardens, chosen mostly for their spring blossom and colourful autumn fruits.

Cockspurthorn fruits and foliage

Cockspurthorn fruits with thorn

Broad-leaved Cockspurthorn flowers

Broad-leaved Cockspurthorn fruits and foliage

Hybrid Cockspurthorn fruits and foliage

Pear-fruited Cockspurthorn fruits and foliage

CHERRIES, PLUMS AND PEACHES *PRUNUS* (FAMILY ROSACEAE)

Trees and large shrubs with attractive flowers, mostly with edible fruits. Single style in each flower results in just one seed, or stone, inside each fleshy fruit, unlike the number of seeds in apples or pears, for example. Many are cultivated for their attractive spring flowers, still more have been domesticated for their edible fruits. Wild species occur in a range of habitats; many are grown in gardens and commercial orchards.

Peach fruit

Peach leaf

Peach *Prunus persica* (Rosaceae) 6m

Small, bushy and rounded deciduous tree. BARK Dark brown. BRANCHES Straight, with smooth, reddish, angular twigs. LEAVES Alternate, lanceolate, finely toothed, often creased into a V-shape. REPRODUCTIVE PARTS Pink flowers, to 4cm across, usually solitary; opening at same time as leaf buds; yellow-tipped anthers. Fruit is familiar peach, to 8cm long, rounded and downy, flushed pink; sweet, juicy flesh when ripe. Seed contained inside a woody, thickly ridged 'stone'. STATUS AND DISTRIBUTION Probably native to China but long cultivated elsewhere. Here, it does best in a walled garden. COMMENTS Nectarine is a smooth-skinned variety of Peach, var. *nectarina*.

Almond nut

Almond *Prunus dulcis* (Rosaceae) 8m

Small, open-crowned tree that blossoms early in spring. BARK Blackish, breaking into small oblong plates. BRANCHES Ascending, usually rather spiny with numerous thin twigs but many cultivars regularly branched and lacking spines. LEAVES Alternate, to 13cm long, finely toothed and folded lengthways. REPRODUCTIVE PARTS Pink or white flowers paired, almost sessile; opening before leaves. 5 petals each to 2.5cm long, forming cup-shaped flowers. Fruit about 6cm long, flattened ovoid, covered with velvety green down with a tough fleshy layer below, inside which is the ridged and pitted 'stone' that, when cracked, reveals edible almond seed. STATUS AND DISTRIBUTION Probably native to central and SW Asia and N Africa, but long cultivated for seeds and flowers. In NW Europe, it needs protection from harsh winter weather.

Almond leaf

Apricot *Prunus armeniaca* (Rosaceae) 10m

Small, rounded deciduous tree. BARK Greyish brown with fine fissures. BRANCHES Twisted and dense with smooth reddish twigs. LEAVES Heart-shaped, reddish when first open, later becoming green above and yellowish beneath, on a red petiole with 2 glands near leaf base. REPRODUCTIVE PARTS White or pale pink short-stalked flowers, solitary or paired, opening before leaves. Fruit rounded, to 8cm long, the downy red-tinged skin surrounding rather acid-tasting juicy flesh that becomes sweet only when fully ripe. Stone flattened, elliptical and smooth, with 3 raised lines along one edge. STATUS AND DISTRIBUTION Native of central and E Asia, grown for its edible fruits; requires shelter in NW Europe.

Apricot fruit

Apricot leaf

Peach flower

Peach fruit

Almond bark

Almond, developing fruits and foliage

Almond mature fruits

Apricot fruit

Espalier Apricot

Apricot flower

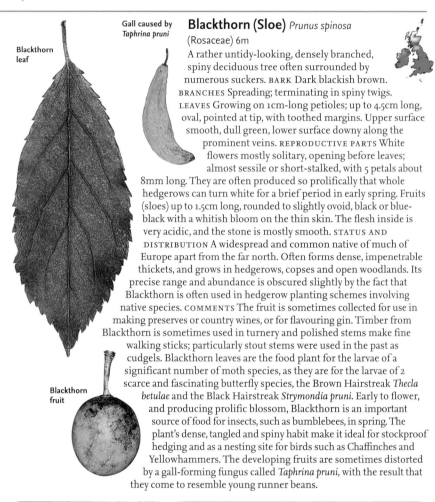

Blackthorn
leaf

Gall caused by
Taphrina pruni

Blackthorn (Sloe) *Prunus spinosa*
(Rosaceae) 6m

A rather untidy-looking, densely branched, spiny deciduous tree often surrounded by numerous suckers. BARK Dark blackish brown. BRANCHES Spreading; terminating in spiny twigs. LEAVES Growing on 1cm-long petioles; up to 4.5cm long, oval, pointed at tip, with toothed margins. Upper surface smooth, dull green, lower surface downy along the prominent veins. REPRODUCTIVE PARTS White flowers mostly solitary, opening before leaves; almost sessile or short-stalked, with 5 petals about 8mm long. They are often produced so prolifically that whole hedgerows can turn white for a brief period in early spring. Fruits (sloes) up to 1.5cm long, rounded to slightly ovoid, black or blue-black with a whitish bloom on the thin skin. The flesh inside is very acidic, and the stone is mostly smooth. STATUS AND DISTRIBUTION A widespread and common native of much of Europe apart from the far north. Often forms dense, impenetrable thickets, and grows in hedgerows, copses and open woodlands. Its precise range and abundance is obscured slightly by the fact that Blackthorn is often used in hedgerow planting schemes involving native species. COMMENTS The fruit is sometimes collected for use in making preserves or country wines, or for flavouring gin. Timber from Blackthorn is sometimes used in turnery and polished stems make fine walking sticks; particularly stout stems were used in the past as cudgels. Blackthorn leaves are the food plant for the larvae of a significant number of moth species, as they are for the larvae of 2 scarce and fascinating butterfly species, the Brown Hairstreak *Thecla betulae* and the Black Hairstreak *Strymondia pruni*. Early to flower, and producing prolific blossom, Blackthorn is an important source of food for insects, such as bumblebees, in spring. The plant's dense, tangled and spiny habit make it ideal for stockproof hedging and as a nesting site for birds such as Chaffinches and Yellowhammers. The developing fruits are sometimes distorted by a gall-forming fungus called *Taphrina pruni*, with the result that they come to resemble young runner beans.

Blackthorn
fruit

Black Hairstreak larva

Blackthorn bark

Blackthorn flowers

Blackthorn fruits

Blackthorn in blossom

Brown Hairstreak

Black Hairstreak

Plum bark

Plum leaf

Bullace leaf

Plum *Prunus domestica* ssp. *domestica* (Rosaceae) 10m

A small deciduous tree. BARK Dull brown, sometimes tinged purple, with deep fissures developing with age. BRANCHES Typically straight branches, usually with no spines (unlike ssp. *insititia* when growing in the wild); twigs brown and smooth. LEAVES Alternate, up to 8cm long, with toothed margins; upper surface smooth green, lower surface downy. REPRODUCTIVE PARTS Flowers mostly white, or sometimes green-tinged, hanging in small clusters of 2–3 on a pedicel 1–2cm long, opening at about same time as leaves in early spring. Fruits up to 7.5cm long, rounded or more often oval, with a smooth skin. Depending on domesticated variety or cultivar, skin may be yellow, red, purple or even green when ripe. The flesh is acidic at first, though never as sharply astringent as that of a sloe, but nearly always becomes sweet when ripe. The flattened oval stone is usually rough and slightly pitted. STATUS AND DISTRIBUTION Probably a hybrid between Blackthorn and Cherry Plum, widely planted throughout Britain and Ireland and naturalised occasionally, nearly always near human habitation. Numerous cultivars exist, each with a particular quality of fruit favoured for its flavour, colour or texture of flesh and suitability for cooking. The timber from Plum is sometimes used in turnery.

Bullace and Damson

Prunus domestica ssp. *insititia* (Rosaceae) 10m

A small deciduous tree. BARK Dull brown, sometimes tinged purple, with deep fissures developing with age. BRANCHES Usually spiny; young twigs and flower stalks downy. LEAVES Alternate, up to 8cm long, with toothed margins, a smooth green upper surface and a downy lower surface. REPRODUCTIVE PARTS Flowers mostly white or sometimes green-tinged, hanging in small clusters of 2–3 on a pedicel 1–2cm long. Opening at about same time as leaves in early spring. Rounded fruits about 5cm long: skin purplish in Damson; Bullace fruits also usually purplish, although some forms yellowish. Flesh rather acidic at first, and slightly astringent (although nothing like as unpalatable as Sloe), but ripening to become sweet and juicy. Oval stone flattened (but less so than in Plum) and usually rough and slightly pitted. STATUS AND DISTRIBUTION Probably a hybrid between Blackthorn and Cherry Plum, widely planted throughout Britain and Ireland and also naturalised (more frequently than Plum), nearly always near human habitation.

Greengage fruit

SIMILAR TREE

The cultivated **Greengage** is now regarded as a distinct subspecies, *P. domestica* ssp. *italica* (10m); fruits yellowish green.

Plum flowers

Bullace flowers

Plum

Plum fruit

Damson fruit

Greengage fruit

Bullace fruit

Wild Cherry (Gean) *Prunus avium* (Rosaceae) 30m

A large deciduous tree with a good tapering bole and a high, domed crown. BARK Reddish brown and shiny, with circular lines of lenticels, peeling horizontally into tough papery strips, and occasionally becoming fissured. BRANCHES Spreading widely and terminating in smooth reddish twigs. LEAVES Up to 15cm long, ovate with a long pointed apex and forward-pointing irregular teeth on the margins. Upper surface smooth and dull, lower surface downy on the veins. Petiole 2–5cm long, with 2 glands near the leaf junction. REPRODUCTIVE PARTS White flowers in long-stalked clusters of 2–6, opening just before the leaves; 5 petals up to 1.5cm long. Fruits up to 2cm long, rounded with a depressed apex and a dark purple or red-black, or more rarely yellow, shiny skin; ripening in early summer. Flesh may be sweet-tasting or rather bitter. Stone rounded and smooth. STATUS AND DISTRIBUTION A widespread native of much of Europe except the far north and east; it is common and widespread throughout the British Isles, except in N Scotland and W Ireland. However, its true native range is often difficult to determine since Wild Cherry is so frequently planted and naturalised. It grows in woods and copses, often in hilly areas. COMMENTS An extremely popular cultivated tree, planted for its spring blossom and timber, and, to a lesser extent, for its fruits. Birds are particularly fond of the ripe fruits and can completely strip a tree within a day or so. The timber, which is hard and strong, is used in turnery and for making furniture, including veneers. Wild Cherry is often used as a stock for grafting other varieties of cherry and is usually cultivated by cuttings. It exists in hundreds of subtly different forms, some of which have cultivar names. 'Plena' is a particularly popular cultivar, with white double flowers that appear later than single-flowered forms; the flower clusters are stalked and pendent.

SIMILAR TREE

Schmitt's Cherry *P.* × *schmittii* (20m) A hybrid between Wild Cherry and Greyleaf Cherry *P. canescens* (an Asian species). Branches slender, mainly erect; bark reddish and peeling into horizontal strips; flowers pink. Planted in urban settings, particularly as a street tree.

Wild Cherry fruit

Cherry turned from Wild Cherry Wood

Wild Cherry leaf

Wild Cherry leaf in autumn

Wild Cherry bark

Wild Cherry

Wild Cherry foliage and fruits

Wild Cherry flowers

Wild Cherry fruits

Cherry Plum leaf

Cherry Plum 'Nigra' leaf

Cherry Plum (Myrobalan Plum)

Prunus cerasifera (Rosaceae) 8m

A small deciduous, rather bushy tree. BARK Dark brown and pitted with rows of white lenticels; fissured in older trees. BRANCHES Many fine and sometimes spiny branches and glossy green twigs. LEAVES To 7cm long, ovate, tapering at base and tip. Margins have numerous rounded teeth and underside has downy veins. Petioles 1cm long, pinkish, grooved. REPRODUCTIVE PARTS White, stalked flowers mostly solitary, opening at about same time as leaves. Fruits to 3.5cm long, rounded, with a smooth red or yellow skin; flesh becoming sweet. Stone smooth, rounded, with a thickened margin. STATUS AND DISTRIBUTION A native of the Balkans, widely planted in Britain for its edible fruits, used for pies and preserves, and also frequently naturalised.

SIMILAR TREE
P. cerasifera 'Pissardii' (8m) Leaves dark reddish green, flowers pale pink or rose-tinted. Often planted in town streets and parks for ornament.

Dwarf Cherry leaf

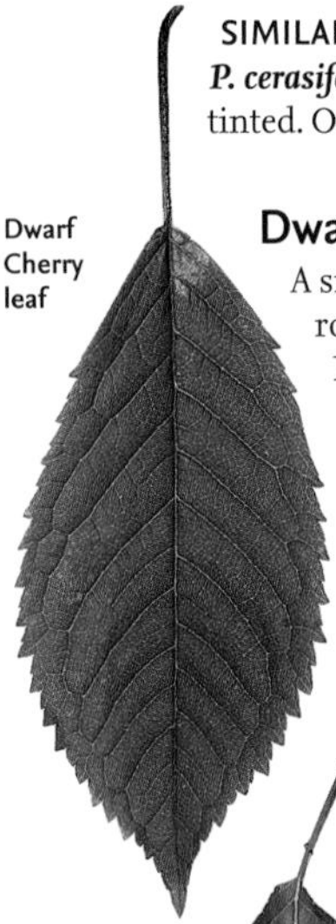

Dwarf Cherry (Sour Cherry) *Prunus cerasus* (Rosaceae) 8m

A small deciduous tree with a very short, branching bole and a rounded shrubby outline, often surrounded by suckers. BARK Reddish brown; twigs smooth. LEAVES To 8cm long, oval to elliptic and sharply pointed at tip, with a tapering base and toothed margin; petioles 1–3cm long. Young leaves slightly downy below, upper surface always smooth and shiny. REPRODUCTIVE PARTS Long-stalked white flowers, usually opening just before leaves in April–May, in clusters of 2–6. Fruits to 1.8cm long, rounded with a slightly depressed apex; usually bright red or blackish red. Flesh soft, tasting acidic; stone rounded and smooth. STATUS AND DISTRIBUTION Native of SW Asia, but widely cultivated for its fruit, which is used mainly in preserves, when it loses much of its acidity. Planted and widely naturalised in Britain, usually by suckering and not bird-sown.

Sargent's Cherry leaf

Sargent's Cherry *Prunus sargentii* (Rosaceae) 13m

Open, spreading tree. BARK Purple-brown, rather glossy and ringed with horizontal bands of lenticels. BRANCHES Ascending or spreading; twigs dark red, thin and smooth. LEAVES To 15cm long, ovate with a long-pointed tip. Margins sharply toothed; smooth on both surfaces. Petiole 2–4cm long. REPRODUCTIVE PARTS Flowers pale pink, on 1–2cm stalks in clusters of 2–4, opening just before leaves, usually in April. Petals up to 2cm long. Fruits (rarely seen here) ovoid, to 1.1cm long, and dark crimson. STATUS AND DISTRIBUTION Native of Japan and Sakhalin; widely planted in Britain. COMMENTS Usually grafted on to a stock of Wild Cherry. Popular as a garden and street tree, favoured for its early spring blossom and rich autumn colours.

Cherry Plum blossom

Cherry Plum flowers

Cherry Plum 'Nigra' flowers

Cherry Plum fruits and foliage

Cherry Plum fruit

Dwarf Cherry fruit

Sargent's Cherry flowers

Japanese Cherry *Prunus serrulata* (Rosaceae) 15m

A small to medium-sized deciduous tree. BARK Purple-brown, ringed by horizontal lines of prominent lenticels. BRANCHES Ascending, usually fanning out from the bole and terminating in smooth twigs. LEAVES Up to 20cm long, ovate and drawn out to a long, tapering tip; margin sharply toothed and petiole smooth, to 4cm long, with up to 4 red glands near the base. REPRODUCTIVE PARTS White or pink flowers in clusters of 2–4, opening just before leaves; in some cultivars flowers are borne on 8cm-long stalks, may have notched petals, and vary in shade from pure white to deep pink. Fruits round, to 7mm long and deep purple-crimson; seldom developing in cultivated trees. STATUS AND DISTRIBUTION Probably native to China, then introduced to Japan at a very early date, and subsequently brought to Britain, where it is now a very popular garden tree. Centuries of breeding and selection have made modern trees different from their wild ancestors, which are rarely seen.

Japanese Cherry leaf

SIMILAR TREES

Originally bred in Japan and ancient in origin, there are many cultivated forms of cherry that are popular elsewhere as garden trees, and that are often collectively referred to as 'Japanese Cherries'; typically they are known only by their cultivar names. Popular cultivars of *Prunus* include: 'Kanzan' with magenta buds and pink flowers; 'Shirofugen' with pink buds and white double flowers; 'Shirotae' with large, white flowers; 'Pink Perfection' with pink double flowers; and 'Tai Haku' (Great White Cherry), with very large white flowers.

Saint Lucie Cherry *Prunus mahaleb* (Rosaceae) 12m

Often little more than a spreading shrub, but sometimes a small tree. BARK Greyish brown and ringed with brown lenticels. BRANCHES Spreading, the young twigs covered with short greyish hairs and often slightly weeping at the tips. LEAVES Alternate, up to 7cm long and almost rounded, with a short point at the tip and a rounded or nearly heart-shaped base. Margin finely toothed, upper surface glossy and lower surface finely downy. REPRODUCTIVE PARTS White, scented flowers in April–May, in clusters of 3–10 in groups of racemes at the end of leafy shoots. The 5 petals are about 8mm long and surround the yellowish anthers. Fruit 0.6–1cm long; an ovate or rounded red berry, ripening black, with bitter-tasting flesh surrounding a smooth rounded stone. STATUS AND DISTRIBUTION A native of central and S Europe, growing in woodland glades and thickets. Planted in Britain for ornament and naturalised occasionally. Several cultivated varieties exist and are grown for ornament in N Europe.

Saint Lucie Cherry leaf

Japanese Cherry flowers

Japanese Cherry

'Japanese Cherry' *Prunus* 'Kanzan' flowers

'Japanese Cherry' *Prunus* 'Shirofugen' flowers

Japanese Cherry autumn colours

Saint Lucie Cherry flowers and foliage

Saint Lucie Cherry fruit and foliage

Spring Cherry *Prunus × subhirtella* (Rosaceae) 20m

Spring
Cherry
leaf

Densely crowned deciduous tree. BARK Greyish brown. BRANCHES Slender, with many downy, crimson twigs. LEAVES To 6cm long, ovate to lanceolate with a long-pointed tip and irregularly toothed margin; veins downy below. Petiole 1cm long, crimson, downy. REPRODUCTIVE PARTS Pinkish-white, short-stalked flowers, opening just before leaves in March or April; petals about 1cm long and notched. Fruits purplish black, rounded, but seldom produced. STATUS AND DISTRIBUTION Native of Japan, commonly planted as a street and garden tree in Britain and Ireland. Various cultivars have been developed, some with a weeping habit, some with double flowers; usually grafted on to stocks of Wild Cherry. COMMENT Winter Cherry *P. × subhirtella* 'Autumnalis' is a common cultivar, unusual in that it flowers throughout winter, from October to April.

Tibetan Cherry *Prunus serrula* (Rosaceae) 15m

Tibetan
Cherry
leaf

Resembles Spring Cherry. BARK Deep purple, peeling to reveal a rich and glossy mahogany-coloured inner layer; often rubbed smooth by passers-by. BRANCHES Spread widely; often pruned to display the bark, the principal decorative feature of this tree. LEAVES To 12cm long, lanceolate and pointed. REPRODUCTIVE PARTS White flowers, opening with leaves, in April–May. Bright red fruits, about 1cm long, sometimes form. STATUS AND DISTRIBUTION Native to China, popular in British gardens for its bark in winter.

Yoshino Cherry *Prunus × yedoensis* (Rosaceae) 15m

Yoshino
Cherry
leaf

A spreading tree, similar to Japanese Cherry. BARK Greyish, with brown lenticel bands. BRANCHES Slightly weeping at tips; young twigs downy. LEAVES Ovate, to 20cm long, with a long, tapering tip and a toothed margin that is 'whiskered' towards the tip. REPRODUCTIVE PARTS Flowers appearing before leaves, in clusters of 5–6 on 2cm pedicels; pale pink petals deeply notched. STATUS AND DISTRIBUTION A hybrid first seen in Japan, now popular as a British street tree.

OTHER HYBRID CHERRIES

Hybrid flowering
cherry leaf

Flowering cherries are popular ornamental trees because of their spring blossom. Numerous cultivars have been developed. Many are favoured in municipal gardens and as street trees. The following are particularly popular cultivars: *Prunus* 'Umineko', with upright to spreading branches, ovate leaves that are toothed and long-tipped, and white flowers that appear just as the leaves are bursting; *Prunus* 'Pandora', with upright and spreading branches, oval and toothed leaves and dull pink flowers; *Prunus* 'Spire', with an extremely erect habit, broadly oval leaves with a slender tip and toothed margins, and dull pink flowers; *Prunus* 'Accolade', with a densely spreading habit, oval to elliptical leaves with toothed margins, and bright pink flowers.

'Autumnalis Rosea' flowers

Spring Cherry flowers

Winter Cherry flowers

Spring Cherry blossom

Spring Cherry blossom

Tibetan Cherry bark

Yoshino Cherry bark

Yoshino Cherry blossom

Bird Cherry *Prunus padus* (Rosaceae) 17m

A deciduous tree, rather conical when young but domed and spreading with age. BARK Smooth, dark and greyish brown, with a strong, unpleasant smell if rubbed. BRANCHES Thin and mostly ascending, terminating in twigs that are smooth, but finely downy when young. LEAVES Alternate and tough, with a dark green upper surface and slightly blue-green underside, elliptical to elongate and up to 10cm long, finely toothed on the margins and tapering at the tip. REPRODUCTIVE PARTS White flowers opening after leaves and growing in 15cm-long spikes that may be pendulous or ascending; they are composed of up to 35 5-petalled flowers, and smell of almonds, the fragrance seeming sickly to some people. Fruits up to 8mm long, shiny black and sour-tasting, rather like sloes. STATUS AND DISTRIBUTION A widespread native across Europe, including Britain and Ireland, where its natural range extends S to Wales and E Anglia; here it is usually found in limestone areas, on stream sides and in damp woods and hedgerows. Bird Cherry does not sucker and so trees are usually encountered in relative isolation from one another. Elsewhere it has been planted and becomes naturalised occasionally; it is favoured for its ornamental value as it is hardy, its blossom makes a good show for a short time in spring, and its leaves are colourful in early autumn. COMMENTS In summer and autumn, birds such as thrushes eagerly consume the fruits, despite their bitter (to our palates) taste. In the past, the bark was used to make a restorative infusion for stomach upsets and as a remedy for other ailments. The timber is hard and reddish, and can be used for carving and turnery.

Bird Cherry leaf

Bird Cherry leaf in autumn

Rum Cherry leaf

Rum Cherry (Black Cherry)

Prunus serotina (Rosaceae) 22m

A spreading deciduous tree with a stout trunk. BARK Greyish, peeling away in strips and fissured in older trees; a strange, bitter smell is released if the bark is damaged. BRANCHES Spreading and dense, the outer extremes sometimes weeping. LEAVES Larger than those of Bird Cherry at up to 14cm long, shiny above and with fine forward-pointing teeth on the margin; the midrib on the underside has patches of hairs along it, helping to separate this species from other similar cherries. REPRODUCTIVE PARTS Flowers very similar to Bird Cherry, but spike may contain fewer than 30 flowers; pedicels shorter and white petals toothed at margins. Black fruits containing bitter-tasting flesh and a rounded smooth stone. STATUS AND DISTRIBUTION A native of North America, planted for timber and ornament in much of Europe, including Britain and Ireland, and naturalised in many places, including S England.

Bird Cherry flowers and foliage

Bird Cherry fruits

Bird Cherry bark

Bird Cherry growing in Ash woodland

Rum Cherry bark

Rum Cherry flowers and foliage

Rum Cherry fruits and foliage

Portugal Laurel leaf

Portugal Laurel *Prunus lusitanica* (Rosaceae) 8m

A small, spreading evergreen tree, or usually a shrub. BARK Smooth or occasionally flaking, and very dark grey to black. BRANCHES Widely spreading; twigs reddish and smooth. LEAVES Dark green, glossy and slightly leathery, and up to 13cm long. Lanceolate to elliptical and tapering at the tip, with a rounded base and a toothed margin. Reddish petiole about 2cm long. REPRODUCTIVE PARTS White flowers in long tapering spikes, up to 26cm long and composed of about 100 strongly scented flowers. They grow out of the leaf axils and exceed the length of the leaves, and are usually pendent. Fruits up to 1.3cm long, ovoid or rounded with a tapering tip, purplish black when ripe and containing a smooth, rounded stone with a ridged margin. STATUS AND DISTRIBUTION A native of Portugal, Spain and SW France, but frequently planted elsewhere in the milder parts of W Europe, including Britain and Ireland, for ornament and as a hedgerow shrub or windbreak; it can survive regular clipping. It is now widely naturalised in Britain, particularly in the south, spreading by layering and by self-sown seeds.

Cherry Laurel *Prunus laurocerasus* (Rosaceae) 8m

An evergreen shrub or small, spreading tree. BARK Dark greyish brown, pitted with numerous lenticels. BRANCHES Dense, with smooth pale green twigs. LEAVES Leathery; up to 20cm long and 6cm wide, oblong with a short-pointed tip and rounded or tapering base. Margin smooth with just a few very small teeth at intervals. REPRODUCTIVE PARTS Fragrant white flowers in an erect spike about 13cm long, which is about the same length as the leaf plus the petiole (flower spike of Portugal Laurel greatly exceeds leaf in length). Fruits rounded and green at first, turning red and then finally blackish purple. They contain a smooth rounded stone with a slightly ridged margin. STATUS AND DISTRIBUTION A native of the E Balkans, commonly planted as an ornamental species since the 16th century in S and W Europe; in Britain and Ireland it is often found naturalised and also exists in a number of cultivars. Cherry Laurel produces suckers freely, so it often spreads out from the original planting to produce dense thickets; it also colonises by means of self-sown seeds. COMMENTS If left unchecked, this colonising species can become a problem, swamping native plants and out-competing native shrubs of similar size. The leaves contain cyanide so maintenance and clearance need to be undertaken with caution.

Cherry Laurel leaf

Portugal Laurel
flowers and foliage

Portugal Laurel blossom

Portugal Laurel fruits

Cherry Laurel unripe fruits

Cherry Laurel
ripe fruits and foliage

Cherry Laurel flowers and foliage

Cherry Laurel flowers and foliage

Judas Tree *Cercis siliquastrum* (Fabaceae) 10m

Judas Tree
leaf

Small, spreading and rather flat-crowned deciduous tree, often with more than one bole. BARK Dark grey. BRANCHES Ascending, spreading near tips, with red-brown buds and twigs. Old trees have more drooping branches. LEAVES Simple, alternate and rounded, sometimes notched at tip and heart-shaped at base; smooth above and bluish green when young, becoming yellow when older; paler and bluer below. REPRODUCTIVE PARTS 5-petalled, pink, pea-like flowers in small short-stalked clusters, opening before leaves and bursting out of bole, large branches and twigs; followed by pods, to 10cm long, slightly constricted around seeds, reddish at first, maturing brown and becoming dehiscent. STATUS AND DISTRIBUTION Native of E Mediterranean, planted in the British Isles as an ornamental tree.

Honey Locust
leaf

Honey Locust *Gleditsia triacanthos* (Fabaceae) 45m

Tall deciduous tree with a high, domed crown; bole, branches and twigs are spiny. BARK Greyish purple. BRANCHES Mainly level, with curled twigs. LEAVES Alternate, either pinnate, with up to 18 pairs of leaflets 2–3cm long, or bipinnate, with up to 14 pairs of leaflets no more than 2cm long. Leaf axes end in spines. REPRODUCTIVE PARTS Tiny flowers in June, no more than 3mm long; may be male, female or both, growing in compact clusters in leaf axils; greenish-white oval petals number from 3 to 5. Flattened pods with thickened edges, to 45cm long, twisted or curved and becoming dark brown when ripe. STATUS AND DISTRIBUTION Native of the Mississippi basin of North America, planted in Britain for ornament, and often seen in the golden-leaved cultivar 'Sunburst'. COMMENTS Cultivars lacking spines are occasionally seen as specimen trees.

Laburnum *Laburnum anagyroides* (Fabaceae) 7m

Deciduous tree with narrow, sparse crown and slender bole. BARK Smooth, greenish brown, marked with blemishes. BRANCHES Often slightly pendulous; shoots grey-green with long, silky, clinging hairs. LEAVES Alternate, divided into 3, each leaflet to 8cm long, elliptic and blunt-pointed at tip, on a petiole 2–6cm long; hairy below when young. REPRODUCTIVE PARTS Copious yellow, fragrant, pea-like flowers in early summer, in pendulous racemes 10–30cm long. Pods, to 6cm long, have smooth blackish-brown, dry outer skin. They persist on the tree, twisting open to reveal pale inner skin and dark seeds. STATUS AND DISTRIBUTION Native of S and central Europe, planted in the British Isles for ornament; sometimes naturalised. COMMENTS Seeds very poisonous.

Laburnum leaf

SIMILAR TREES

Scottish Laburnum *L. alpinum* (12m) Shoots and leaves hairless; flowers in racemes to 40cm long.

Voss's
Laburnum
leaf

Voss's Laburnum *L. × watereri* 'Vossii' (11m) A hybrid between Laburnum and Scottish Laburnum, with the early flowers of the first and the longer, more densely packed racemes of the second. Where laburnums are planted for ornament they are likely to be of this type: it has good hybrid vigour and makes a finer, longer-lived tree.

Scottish Laburnum
leaf

Judas Tree flowers and foliage

Honey Locust flowers and foliage

Honey Locust foliage

Honey Locust spines and foliage

Laburnum flowers

Voss's Laburnum flowers

Voss's Laburnum flowers

Mimosa leaf

Mimosa *Acacia dealbata* (Fabaceae) 30m

Medium-sized tree. Twigs, shoots and foliage are covered by silvery white hairs. BARK Smooth, greenish grey, blackening with age. BRANCHES Upright. LEAVES Fern-like, tripinnate; leaflets to 5mm long. REPRODUCTIVE PARTS Tiny yellow flowers, in small globular heads of 30–40 flowers, on long racemes of 20–30 heads. Pods flattened, to 10cm long; not constricted between seeds. STATUS AND DISTRIBUTION Native of Australia, grown for ornament in the British Isles.

False Acacia *Robinia pseudoacacia* (Fabaceae) 30m

Medium-sized, open-crowned tree. BARK Spirally ridged. BRANCHES Easily snapped. LEAVES Alternate, to 20cm long, pinnate, with 3–10 pairs of oval yellowish-green leaflets; petiole has 2 woody, basal stipules, each leaflet has a small stipule at petiole base. REPRODUCTIVE PARTS Fragrant, white pea-like flowers in dense, hanging clusters, to 20cm long. Pods smooth, to 10cm long. STATUS AND DISTRIBUTION Native to USA, planted in Britain and naturalised. COMMENTS Cultivar 'Frisia' (Golden Robinia), a popular smaller tree, has golden-yellow foliage in summer that turns orange in autumn.

False Acacia leaf

Pink Siris *Albizia julibrissin* (Fabaceae) 13m

Small tree. BARK Smooth and grey. BRANCHES Spreading. LEAVES Pinnate, to 40cm long; each leaf has up to 25 pinnae, each pinna with about 35–50 curved, ovate leaflets, to 1.5cm long, green above, white below. REPRODUCTIVE PARTS Pinkish-orange plume-like flowers on branched, hairy stalks; pods brown, to 15cm long, constricted between seeds. STATUS AND DISTRIBUTION Native of Asia, planted occasionally in Britain.

Pink Siris leaf

Tree of Heaven leaf

Tree of Heaven

Ailanthus altissima (Simaroubaceae) 20m

Vigorous, suckering tree. BARK Smooth and grey at first, pale and scaly with age. BRANCHES Thick, mostly upright; twigs ending in tiny, ovoid, scarlet buds. LEAVES Alternate, pinnate, to 60cm long, with up to 25 pointed leaflets 7–12cm long; deep red at first, shiny green in summer. REPRODUCTIVE PARTS Greenish flowers in fairly open spikes on trees of separate sexes. Fruits reddish, winged and twisted seeds about 3cm long. STATUS AND DISTRIBUTION Native of China, widely planted in the British Isles.

Stag's-horn Sumach *Rhus typhina* (Anacardiaceae) 10m

Small, spreading tree. BARK Brown. BRANCHES Downy. LEAVES Alternate, pinnate, with up to 29 leaflets, each leaflet up to 12cm long, coarsely toothed; fiery autumn colours. REPRODUCTIVE PARTS Tiny flowers borne on trees of separate sexes; male flowers greenish, female flowers red, in dense conical clusters, to 20cm long, at tips of twigs. Fruits resemble small nuts. STATUS AND DISTRIBUTION Native of North America, widely planted in Britain and sometimes naturalised.

Mimosa foliage

Mimosa flowering tree

Pink Siris flower

False Acacia flowers and foliage

False Acacia

tag's-horn Sumach autumn colours

Tree of Heaven
flowers and foliage

Tree of Heaven

Field Maple *Acer campestre* (Aceraceae) 26m

A medium-sized deciduous tree with a rounded crown and a twisted bole. Rather variable in appearance, much depending on its habitat. BARK Grey-brown and fissured with a slightly corky texture. BRANCHES Much divided and dense, sometimes almost impenetrably so, when pruned or cut regularly. The shoots are brown, sometimes covered with fine hairs and often developing wings, especially in trees regularly pruned back in hedgerows. LEAVES Opposite, up to 12cm long and usually strongly 3-lobed. The lobes themselves often have lobed margins and tufts of hair in the axils of the veins on the underside. Newly opened leaves have a pinkish tinge at first, becoming dark green and rather leathery later. They turn bright yellow then reddish brown in autumn, producing an excellent splash of colour in hedgerows. REPRODUCTIVE PARTS Yellowish-green flowers in small, open, erect clusters; there are 5 sepals and 5 petals. Male and female flowers occur together, opening at about the same time as the leaves in April–May. Winged fruits in bunches of 4; wings horizontal and greenish tinged with varying amounts of red. The hard seeds are at the base. The wings allow the seeds to be carried a considerable distance from the parent tree by the wind, aiding dispersal and colonisation. STATUS AND DISTRIBUTION A widespread and common native tree in N Europe, including parts of Britain, occurring in woods and hedgerows; its precise natural range and abundance is obscured in some areas because it is widely used in planting schemes involving native species, being favoured for its beautiful autumn colours. This is the only *Acer* species that occurs naturally in Britain, its native range covering much of England and Wales; it has been introduced to Scotland and Ireland. It thrives on calcareous soils, doing particularly well on slopes of chalk downs in SE England, often colonising newly cleared or disturbed ground. It is seldom found on acid soils or in waterlogged conditions. COMMENTS Field Maple timber is buffish brown and, in the past, was used for turnery and carving. Fallen seeds are an important source of food for Wood Mice and Bank Voles, and the leaves are eaten by the larvae of a number of moth species. Nail galls sometimes appear on the leaves of young trees, caused by the Sycamore Gall Mite *Aculops acericola*. A range of cultivars exists, all with subtly different leaf colours from the wild tree; cultivar leaf colours range from variegated green and white to deep red and pale pink.

Field Maple leaf

Field Maple leaf in autumn

Field Maple seeds

Field Maple bark

Field Maple autumn colour

Field Maple

Field Maple autumn foliage

Field Maple flowers and foliage

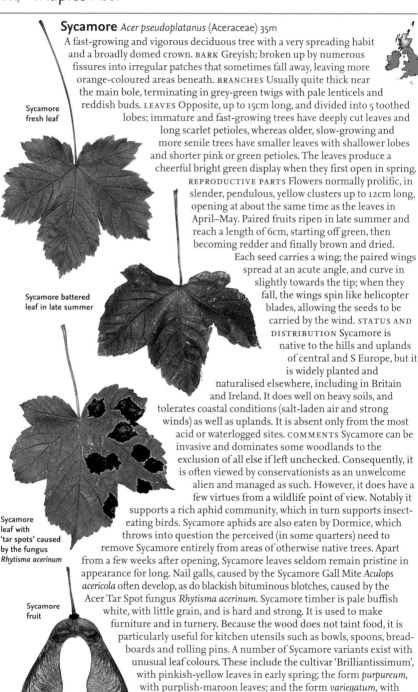

Sycamore *Acer pseudoplatanus* (Aceraceae) 35m

A fast-growing and vigorous deciduous tree with a very spreading habit and a broadly domed crown. BARK Greyish; broken up by numerous fissures into irregular patches that sometimes fall away, leaving more orange-coloured areas beneath. BRANCHES Usually quite thick near the main bole, terminating in grey-green twigs with pale lenticels and reddish buds. LEAVES Opposite, up to 15cm long, and divided into 5 toothed lobes; immature and fast-growing trees have deeply cut leaves and long scarlet petioles, whereas older, slow-growing and more senile trees have smaller leaves with shallower lobes and shorter pink or green petioles. The leaves produce a cheerful bright green display when they first open in spring. REPRODUCTIVE PARTS Flowers normally prolific, in slender, pendulous, yellow clusters up to 12cm long, opening at about the same time as the leaves in April–May. Paired fruits ripen in late summer and reach a length of 6cm, starting off green, then becoming redder and finally brown and dried. Each seed carries a wing; the paired wings spread at an acute angle, and curve in slightly towards the tip; when they fall, the wings spin like helicopter blades, allowing the seeds to be carried by the wind. STATUS AND DISTRIBUTION Sycamore is native to the hills and uplands of central and S Europe, but it is widely planted and naturalised elsewhere, including in Britain and Ireland. It does well on heavy soils, and tolerates coastal conditions (salt-laden air and strong winds) as well as uplands. It is absent only from the most acid or waterlogged sites. COMMENTS Sycamore can be invasive and dominates some woodlands to the exclusion of all else if left unchecked. Consequently, it is often viewed by conservationists as an unwelcome alien and managed as such. However, it does have a few virtues from a wildlife point of view. Notably it supports a rich aphid community, which in turn supports insect-eating birds. Sycamore aphids are also eaten by Dormice, which throws into question the perceived (in some quarters) need to remove Sycamore entirely from areas of otherwise native trees. Apart from a few weeks after opening, Sycamore leaves seldom remain pristine in appearance for long. Nail galls, caused by the Sycamore Gall Mite *Aculops acericola* often develop, as do blackish bituminous blotches, caused by the Acer Tar Spot fungus *Rhytisma acerinum*. Sycamore timber is pale buffish white, with little grain, and is hard and strong. It is used to make furniture and in turnery. Because the wood does not taint food, it is particularly useful for kitchen utensils such as bowls, spoons, bread-boards and rolling pins. A number of Sycamore variants exist with unusual leaf colours. These include the cultivar 'Brilliantissimum', with pinkish-yellow leaves in early spring; the form *purpureum*, with purplish-maroon leaves; and the form *variegatum*, with variegated leaves.

Sycamore
fresh leaf

Sycamore battered
leaf in late summer

Sycamore
leaf with
'tar spots' caused
by the fungus
Rhytisma acerinum

Sycamore
fruit

Sycamore bark

Sycamore fruit and foliage

Sycamore tree in hedgerow

Sycamore fruit and foliage

Sycamore leaves

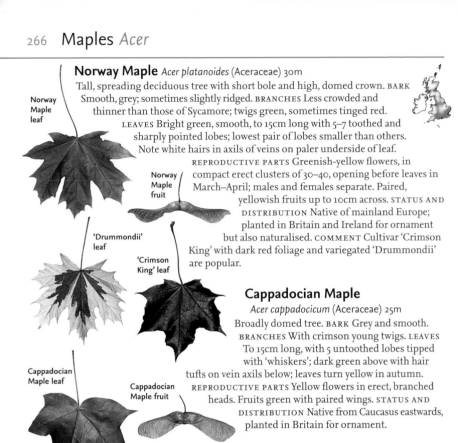

Norway Maple *Acer platanoides* (Aceraceae) 30m

Norway Maple leaf

Tall, spreading deciduous tree with short bole and high, domed crown. BARK Smooth, grey; sometimes slightly ridged. BRANCHES Less crowded and thinner than those of Sycamore; twigs green, sometimes tinged red. LEAVES Bright green, smooth, to 15cm long with 5–7 toothed and sharply pointed lobes; lowest pair of lobes smaller than others. Note white hairs in axils of veins on paler underside of leaf. REPRODUCTIVE PARTS Greenish-yellow flowers, in compact erect clusters of 30–40, opening before leaves in March–April; males and females separate. Paired, yellowish fruits up to 10cm across. STATUS AND DISTRIBUTION Native of mainland Europe; planted in Britain and Ireland for ornament but also naturalised. COMMENT Cultivar 'Crimson King' with dark red foliage and variegated 'Drummondii' are popular.

Norway Maple fruit

'Drummondii' leaf

'Crimson King' leaf

Cappadocian Maple

Acer cappadocicum (Aceraceae) 25m

Broadly domed tree. BARK Grey and smooth. BRANCHES With crimson young twigs. LEAVES To 15cm long, with 5 untoothed lobes tipped with 'whiskers'; dark green above with hair tufts on vein axils below; leaves turn yellow in autumn. REPRODUCTIVE PARTS Yellow flowers in erect, branched heads. Fruits green with paired wings. STATUS AND DISTRIBUTION Native from Caucasus eastwards, planted in Britain for ornament.

Cappadocian Maple leaf

Cappadocian Maple fruit

Montpelier Maple *Acer monspessulanum* (Aceraceae) 15m

Small deciduous tree with a neatly domed crown. BARK Blackish or grey and fissured. BRANCHES With smooth, thin, brown twigs terminating in small, ovoid orange-brown buds. LEAVES Leathery, to 8cm long, with 3 spreading lobes and entire margins; shiny dark green above and bluish below with a few tufts of hairs in axils of lower veins; petiole of similar length to leaf and orange-tinted. Leaves fresh green in spring, but dark in summer, remaining on tree until well into autumn. REPRODUCTIVE PARTS Yellowish-green flowers after leaves, in small clusters on long, slender pedicels; upright at first but pendent later. Red-tinged fruits about 1.2cm long, with parallel or overlapping wings. STATUS AND DISTRIBUTION Native of S Europe, planted in Britain for ornament.

Montpellier Maple fruit

Montpellier Maple leaf

Cretan Maple *Acer sempervirens* (Aceraceae) 12m

Shrub or small, compact evergreen tree. BARK Smooth and grey with lighter patches. BRANCHES Tangled and twisted with shiny brown twigs. LEAVES Opposite, to 5cm long and often 3-lobed, but sometimes irregular or simple with untoothed margins. REPRODUCTIVE PARTS Greenish flowers opening in April, in small erect clusters. Fruits greenish, red-winged, with wings parallel or slightly divergent. STATUS AND DISTRIBUTION Native of Crete and Greece, planted in Britain occasionally.

Cretan Maple fruit

Cretan Maple leaf

Norway Maple bark

Norway Maple flowers

Norway Maple autumn colours and fruits

Cappadocian Maple

Cappadocian Maple foliage

Montpelier Maple foliage

Cretan Maple fruits and foliage

Sugar Maple leaf

Sugar Maple *Acer saccharum* (Aceraceae) 26m

Similar to Norway Maple. BARK With large fissures, and falling away in shreds in older trees. BRANCHES Upright to spreading. LEAVES 13cm long, lobed, but teeth on lobes rounded, not drawn out into a fine point as in Norway Maple; hairs in vein axils below. REPRODUCTIVE PARTS Pendulous yellow-green flowers small and lacking petals; opening in spring with leaves. STATUS AND DISTRIBUTION Native of E North America, planted in Britain for its autumn colours.

Silver Maple *Acer saccharinum* (Aceraceae) 30m

Broadly columnar tree with spreading crown; suckers freely. BARK Smooth, greyish, but scaly with age. BRANCHES Numerous, slender and ascending with pendulous brownish twigs. LEAVES To 16cm long, deeply divided into 5 lobes with irregularly toothed margins, orange or red-tinted at first, green above later, but with silvery hairs below; petiole usually pink-tinged. REPRODUCTIVE PARTS Yellowish-green flowers (no petals) in small, short-stalked clusters of separate sexes in spring. Fruits green, then brown, about 6cm long, with diverging wings and prominent veins. STATUS AND DISTRIBUTION Native of E North America, planted in Britain for ornament. COMMENTS The Cut-leaved Silver Maple, f. *laciniatum*, is frequently found in city squares.

Silver Maple leaf

Sugar Maple fruit

Red Maple *Acer rubrum* (Aceraceae) 23m

Fast-growing, spreading tree with an irregular crown. BARK Grey and smooth. BRANCHES Mostly ascending, but arching outwards. LEAVES To 10cm long and almost as wide, with 3–5 toothed lobes less than half the leaf width; red-tinged above at first, greener later, and silvery below, with a red petiole. Turning various shades of red and yellow in autumn. REPRODUCTIVE PARTS Small red flowers in dense clusters on thin pedicels, opening in spring before leaves. Males and females separate. Fruits bright red, winged, about 1cm long, the wings diverging at a narrow angle. STATUS AND DISTRIBUTION Native of E North America, usually growing in damp habitats; grown in Britain for its autumn foliage.

Red Maple leaf

Moosewood *Acer pensylvanicum* (Aceraceae) 14m

One of the so-called 'snakebark maples'. BARK Green, vertically striped with reddish brown or white; becoming greyer with age. BRANCHES Mainly upright. LEAVES To 15cm long and about same width, with 3 triangular forward-pointing lobes that taper to slender points; central lobe longest. Rich yellow-green in summer, with a smooth upper surface and a hairy lower surface when first open; turning deep yellow in autumn. REPRODUCTIVE PARTS Small yellow-green flowers, in pendulous racemes, appearing in spring with leaves. Greenish fruits about 2.5cm long with downcurved wings. STATUS AND DISTRIBUTION Native of North America, planted in Britain for its autumn colours.

Moosewood leaf

Moosewood fruit

Sugar Maple foliage

Silver Maple leaf upperside

Silver Maple leaf underside

Red Maple autumn colour

Red Maple autumn colour

Moosewood, young tree bark

Moosewood, mature tree bark

Moosewood leaf

NOTE Several maples, from China and Japan, have similarly attractive bark to Moosewood (*see* p. 268). Of these, Père David's and Red Snakebark Maples are the most regularly encountered.

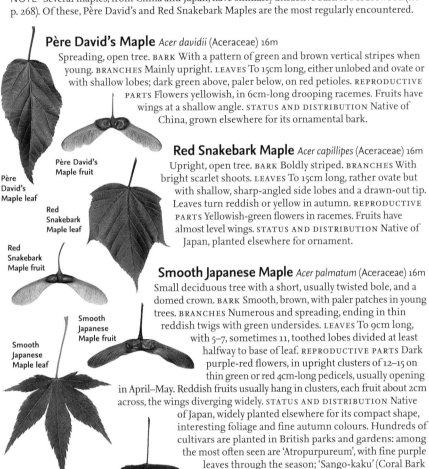

Père David's Maple *Acer davidii* (Aceraceae) 16m

Spreading, open tree. BARK With a pattern of green and brown vertical stripes when young. BRANCHES Mainly upright. LEAVES To 15cm long, either unlobed and ovate or with shallow lobes; dark green above, paler below, on red petioles. REPRODUCTIVE PARTS Flowers yellowish, in 6cm-long drooping racemes. Fruits have wings at a shallow angle. STATUS AND DISTRIBUTION Native of China, grown elsewhere for its ornamental bark.

Père David's Maple fruit

Père David's Maple leaf

Red Snakebark Maple *Acer capillipes* (Aceraceae) 16m

Upright, open tree. BARK Boldly striped. BRANCHES With bright scarlet shoots. LEAVES To 15cm long, rather ovate but with shallow, sharp-angled side lobes and a drawn-out tip. Leaves turn reddish or yellow in autumn. REPRODUCTIVE PARTS Yellowish-green flowers in racemes. Fruits have almost level wings. STATUS AND DISTRIBUTION Native of Japan, planted elsewhere for ornament.

Red Snakebark Maple leaf

Red Snakebark Maple fruit

Smooth Japanese Maple *Acer palmatum* (Aceraceae) 16m

Small deciduous tree with a short, usually twisted bole, and a domed crown. BARK Smooth, brown, with paler patches in young trees. BRANCHES Numerous and spreading, ending in thin reddish twigs with green undersides. LEAVES To 9cm long, with 5–7, sometimes 11, toothed lobes divided at least halfway to base of leaf. REPRODUCTIVE PARTS Dark purple-red flowers, in upright clusters of 12–15 on thin green or red 4cm-long pedicels, usually opening in April–May. Reddish fruits usually hang in clusters, each fruit about 2cm across, the wings diverging widely. STATUS AND DISTRIBUTION Native of Japan, widely planted elsewhere for its compact shape, interesting foliage and fine autumn colours. Hundreds of cultivars are planted in British parks and gardens: among the most often seen are 'Atropurpureum', with fine purple leaves through the season; 'Sango-kaku' (Coral Bark Maple), with pinkish-red shoots and golden leaves; and 'Osakazuki', with brilliant scarlet autumn leaves and fruits.

Smooth Japanese Maple fruit

Smooth Japanese Maple leaf

Downy Japanese Maple leaf

'Osakazuki' leaf in autumn

Downy Japanese Maple

Acer japonicum (Aceraceae) 14m

Similar to Smooth Japanese Maple, but bole often even shorter; note also differences between the leaves. BARK Grey and smooth. BRANCHES Upright and sinuous. LEAVES Hairy when young, with veins remaining hairy through the season. Leaves lobed, but divided less than halfway to base, with forward-pointing teeth on margins. REPRODUCTIVE PARTS Purple flowers in long-stalked, pendulous clusters, opening just before leaves. Paired, winged fruits, to 5cm across, the wings diverging widely; margins hairy at first. STATUS AND DISTRIBUTION Native of Japan, grown in Britain for ornament; cultivar 'Vitifolium' is popular for its red autumn colours.

Downy Japanese Maple fruit

Père David's Maple bark

Red Snakebark Maple bark

Père David's Maple fruits and foliage

Red Snakebark Maple flowers and foliage

Smooth Japanese Maple 'Atropurpureum'

Smooth Japanese Maple autumn colours

Smooth Japanese Maple fruits and foliage

Downy Japanese Maple foliage

Paper-bark Maple *Acer griseum* (Aceraceae) 15m

Paper-bark Maple leaf

Dense and spreading tree. BARK Reddish brown and distinctive, peeling off in thin papery scales. BRANCHES Mainly level. LEAVES Pinnate and divided into 3 blunt-toothed leaflets, each toothed and lobed. REPRODUCTIVE PARTS Small yellow-green flowers in drooping clusters. Pale green winged fruits about 3cm long. STATUS AND DISTRIBUTION Native of China, planted elsewhere in gardens for its ornamental bark.

Paper-bark Maple fruit

Nikko Maple *Acer nikoense* (Aceraceae) 15m

Broadly spreading deciduous tree. BARK Greyish brown and smooth. BRANCHES Mainly level, with blackish buds that have grey hairs on scales. LEAVES Compound, with 3 leaflets, the central one up to 10cm long, the other 2 smaller and unequal at the base. They are mostly green and smooth on the upper surface, but bluish white below with a covering of soft hairs. Leaves turn fiery red in autumn. REPRODUCTIVE PARTS Small yellow flowers, in pendulous clusters of 3, on hairy stalks, opening at about same time as leaves. Green, winged fruits about 5cm long, with wings slightly spreading, but seeds are rarely fertile or fully formed. STATUS AND DISTRIBUTION Native of Japan, and now popular in Britain as an ornamental tree, mostly for its fine autumn colours.

Nikko Maple leaf in autumn

Nikko Maple fruit

SIMILAR TREE

Rough-barked Maple *A. triflorum* (12m) Best recognised by its pale, grey-brown bark that peels in vertical strips. Pinnate leaves composed of 3 sparsely toothed, bristly hairy leaflets that turn a bright orange or red in autumn. Native of NE China and Korea, grown in Britain for its autumn colours.

Ashleaf Maple (Box Elder)

Ashleaf Maple leaf

Ashleaf Maple fruit

Acer negundo (Aceraceae) 20m

Small but vigorous deciduous tree with numerous shoots growing from bole and main branches, giving it a rather crowded and untidy appearance. BARK Smooth in young trees, replaced by darker, shallowly fissured bark in older trees. BRANCHES With green shoots and small buds that have only 2 whitish scales. LEAVES Pinnate, to 15cm long with 3 or sometimes up to 7 irregularly toothed oval leaflets. REPRODUCTIVE PARTS Male and female flowers occur separately, opening in March before leaves. Petals absent; male flowers greenish with prominent red anthers; female flowers greenish yellow and pendent. Brown fruits about 2cm long with wings slightly spreading, remaining on tree after leaves have fallen. STATUS AND DISTRIBUTION Native of E North America, commonly planted as an ornamental tree, and sometimes for shelter; sometimes naturalised. COMMENTS Cultivar 'Variegatum' is now a common street and town park tree. It is recognised by its green and yellow variegated leaves, although it sometimes reverts to the plain green of the original tree; variegated form bears only female flowers.

Ashleaf Maple leaf, variegated form

Paper-bark Maple bark

Paper-bark Maple bark

Paper-bark Maple fruits and foliage

Nikko Maple foliage upperside

Nikko Maple foliage underside

Ashleaf Maple foliage

Horse-chestnut leaf

Horse-chestnut

Aesculus hippocastanum (Hippocastanaceae) 25m
A large deciduous tree with a massive, domed crown. BARK Greyish brown, often flaking away in large scales. BRANCHES Snapping off readily when large; reddish-brown twigs have numerous whitish lenticels. The winter buds are a conspicuous and familiar feature, being shiny brown and sticky, up to 3.5cm long (*see* p. 11). Below each bud is a shield-shaped leaf scar with raised bumps around the edges, fancifully resembling a horseshoe. LEAVES Large, long-stalked and palmate, composed of up to 7 leaflets, each up to 25cm long, the central leaflets being the longest, all of them sharply toothed and elongate-oval. Upper surface mostly smooth, lower surface slightly downy. REPRODUCTIVE PARTS Flowers abundant, often covering the tree in a mass of creamy white panicles, each made up of 40 or more 5-petalled, pink-spotted white flowers, and reaching a length of 30cm. Fruits spiny-cased and rounded, reaching a length of about 6cm and containing a single large round seed ('conker'), or occasionally 2 or 3 flattened seeds, each one bearing a large pale scar. STATUS AND DISTRIBUTION Native of the mountains of the Balkans, but widely planted over

Horse-chestnut fruit

Horse-chestnut seed (conker)

much of Europe, apart from the far north. Long established in Britain and Ireland: Horse-chestnut arrived here in the late 16th century. COMMENTS A variety of explanations exist as to the origins of the Horse-chestnut's English name. The most plausible is that 'horse' refers to the conker's use (in 16th-century Turkey at least) in treating ailing horses; the chestnut part of the name is presumed to relate to the fact that the fruits are superficially similar to those of unrelated Sweet Chestnut. With the tree's medicinal properties in mind, it is interesting to note that an extract from the fruit is used in herbal medicine to treat vascular disorders in humans. The flowers of Horse-chestnut are popular with nectar-seeking insects, in particular bees. The soft, pale timber is of no economic importance.

SIMILAR TREE

Red Horse-chestnut *A.* × *carnea* (20m) A hybrid between Horse-chestnut and Red Buckeye (*see* p. 276), forming a sizeable, domed tree with a gnarled bole and twisted branches. Leaves composed of 5–7 leaflets, each dark green and with toothed margins. Flowers similar to those of Horse-chestnut: sometimes creamy white with yellow blotches at first but turning pink with red blotches. Widely planted in parks and formal gardens.

Red Horse-chestnut fruit

Horse-chestnut bark

Horse-chestnut

Horse-chestnut fruit (conkers)

Horse-chestnut

Red Horse-chestnut flowers

Red Horse-chestnut flowers and foliage

Red Buckeye *Aesculus pavia* (Hippocastanaceae) 5m

Small, spreading deciduous tree with a domed crown. BARK Smooth and dark grey. BRANCHES Level or slightly weeping. LEAVES Palmate, composed of 5 lanceolate, pointed, sharply toothed, short-stalked leaflets; dark glossy green above, turning red in autumn. REPRODUCTIVE PARTS Slender red flowers, to 4cm long, with 4 petals; in erect spikes in early summer. Fruits rounded or pear-shaped, with a smooth brown outer skin enclosing 1 or 2 shiny brown seeds. STATUS AND DISTRIBUTION Native of SE USA, planted elsewhere for ornament.

Red Buckeye leaf

Indian Horse-chestnut

Aesculus indica (Hippocastanaceae) 30m

Large, broadly columnar tree with a thick trunk. Resembles Horse-chestnut, but more graceful, especially in winter outline. BARK Smooth, greyish green or tinged pink. BRANCHES Ascending, but with pendulous twigs and shoots. LEAVES Like those of Horse-chestnut but leaflets are narrower, stalked and finely toothed, to 25cm long; tinged bronze when young, green in summer, turning yellow or orange in autumn. REPRODUCTIVE PARTS Flowers in midsummer, white or pale pink with bright yellow blotches and long stamens extending out of flower; yellow blotch becoming red as flower matures. Flower spikes erect, to 30cm long. Stalked brown fruits pear-shaped and scaly with up to 3 seeds. STATUS AND DISTRIBUTION Native of Himalayas. Planted in Britain occasionally.

Indian Horse-chestnut leaf

Yellow Buckeye (Sweet Buckeye)

Aesculus flava (Hippocastanaceae) 30m

Large domed, deciduous tree. BARK Peeling and scaly grey-brown. BRANCHES Ascending and twisted. LEAVES Palmate with 5 leaflets, each up to 20cm long; turning red early in the autumn. REPRODUCTIVE PARTS 4-petalled yellow flowers in erect spikes about 15cm long, usually opening in late spring or early summer. Smooth, rounded fruits about 6cm across, covered in brown scales on the outside and containing 1 or 2 seeds. STATUS AND DISTRIBUTION Native of E USA, planted in British parks and gardens for its excellent autumn colours.

Yellow Buckeye fruit

Yellow Buckeye leaf in autumn

California Buckeye *Aesculus californica* (Hippocastanaceae) 13m

Broad, spreading and domed deciduous tree. BARK Greyish pink and smooth. BRANCHES Upright to level; buds sticky. LEAVES Small and palmate with 5–7 leaflets, each 5–10cm long and dark glossy green. REPRODUCTIVE PARTS Flowers whitish, sometimes flushed pink, and fragrant; borne in 'candles' in early summer. Fruits have rough, but not spiny, husks. STATUS AND DISTRIBUTION Native of California, planted in Britain occasionally for ornament.

California Buckeye leaf

Red Buckeye flowers and foliage

Indian Horse-chestnut flowers and foliage

Indian Horse-chestnut bark

Yellow Buckeye bark

Yellow Buckeye fruit and foliage

Yellow Buckeye bud

California Buckeye flowers

Holly leaf

Holly, variegated leaf

Holly *Ilex aquifolium* (Aquifoliaceae) 15m

A striking evergreen tree with fine, shiny dark green foliage with very strong prickles. Sometimes only a shrub, but can grow into a handsome conical tree. BARK Smooth silver-grey with fissures and tubercles appearing with age. BRANCHES Sweeping downwards but with tips of younger branches turning up. Shoots and buds green. LEAVES Alternate and up to 12cm long, tough and leathery with a waxy upper surface and a paler lower surface. Margins wavy and spiny, although there may be some variation on any one tree, with leaves from the upper branches of a large tree being much flatter and mostly spineless. REPRODUCTIVE PARTS White flowers about 6mm across and 4-petalled, in small clusters in the leaf axils. Males and females on different trees; male flowers fragrant. Trees that grow in the shaded understorey of a woodland are usually sterile. Fruit a bright red, stalked berry with a thin fleshy layer, up to 12mm long. STATUS AND DISTRIBUTION A native tree of woodlands and more open habitats across most of W and S Europe, and parts of W Asia. It is native to Britain and Ireland, where it is widespread and common, but it is also widely planted as an ornamental tree in parks and gardens. Holly is a common component of the understorey of oak and beech woods. Hedgerow trees are often mutilated in winter as people gather branches for Christmas decorations. COMMENTS Woodlands with a deer population often have Hollies with few lower branches and a distinct browse line – the leaves and shoots are a favourite food in mid-winter when little else is green. Holly berries are popular with thrushes in winter. However, a particularly productive tree is often 'owned' by a resident Mistle Thrush, which will drive off other birds. Holly flower buds are food for the first brood larvae of the Holly Blue butterfly. Leaves are often marked with brown blotches, caused by the larva of the Holly Leaf Miner *Phytomyza ilicis*.

SIMILAR TREE

Highclere Holly *I. × altaclerensis* (20m) A descendant of hybrid crosses between Holly and Madeira Holly *I. perado* (not hardy in Britain). An evergreen tree with a dense columnar habit, spreading branches and a domed crown. Bark purplish grey, twigs greenish or tinged purple. Alternate leaves mostly flat, smaller than those of Holly and not as prickly, with up to 10 small forward-pointing spines on each side. Flowers small, white, 5-petalled, sometimes tinged purple near the base and, as with Holly, males and females on separate trees. Berries bright red, up to 12mm long. In its various cultivar forms, a very popular park and garden tree because of its vigorous habit and resistance to disease and pollution. It grows well in towns and near the sea. Of the numerous cultivars, most have variegated leaves; some look almost completely gold.

Highclere Holly leaves

Holly Blue butterfly

Holly fruits and foliage

Holly flowers and foliage

Holly

Holly, variegated form

Holly Leaf Miner

Highclere Holly
fruits and foliage

Spindle bark

Spindle
leaf

Spindle fruit

Spindle *Euonymus europaeus* (Celastraceae) 6m
Slender, sometimes spreading and rather twiggy
deciduous tree. BARK Smooth and grey, becoming
slightly fissured and pink-tinged as the tree
ages. BRANCHES Numerous, the green twigs
having an angular feel when young but
becoming rounded when older, and terminating in
tiny pointed buds. LEAVES Ovate, to 10cm long, with
a pointed tip and sharply toothed margins; they turn
a rich shade of purple-orange in autumn.
REPRODUCTIVE PARTS Small yellowish-green, 4-
petalled flowers in clusters in leaf axils, opening in
early summer. Fruits are pink capsules about 1.5cm
across and divided into 4 chambers, each containing an
orange seed. STATUS AND DISTRIBUTION Native to
much of Europe except the extreme north and south. It is
native to England, Wales and Ireland, and is found mainly
in hedgerows and copses, especially on lime-rich soils.

SIMILAR TREES
Large-leaved Spindle *E. latifolius* (6m) Very similar to Spindle, but
twigs not as markedly angled, and buds longer and more pointed.
Leaves to 16cm long, elliptical, with finely toothed margins.
Flowers 5-petalled, pink, in lax clusters of 4–12 in leaf axils. Capsule pink, more sharply angled,
divided into 5 chambers each containing a single orange seed. Native to S, central and SE
Europe, but planted in Britain and very occasionally naturalised.
Evergreen Spindle *E. japonicus* (6m) Leaves shiny, evergreen, obovate, to 7cm long. Pink fruits have
4 rounded lobes. Native to Japan, planted in Britain and very occasionally naturalised.

Box bark

Box leaf

Box *Buxus sempervirens* (Buxaceae) 6m
Small, very dense, spreading evergreen tree or
large shrub. BARK Smooth and grey, breaking
into small squares with age. BRANCHES
Numerous; young twigs green, angular and
covered with white hairs. LEAVES Opposite,
ovate to oblong, to 2.5cm long and 1cm across,
with a notched tip; upper surface dark green and
glossy, lower surface paler. REPRODUCTIVE PARTS
Flowers small and green, males with conspicuous
yellow anthers; in same cluster as female flowers,
opening in early spring. Fruit a small, woody,
greenish capsule, about 8mm long, with 3 spreading
spines; ripening to brown, then splitting open to
scatter the hard, shiny black seeds. STATUS AND
DISTRIBUTION Local native of dry, calcareous hills and
slopes in mainland Europe. In Britain, it is thought to be native to certain chalk slopes in
S England but because it is so widely planted its status elsewhere is uncertain. COMMENTS
Several different cultivars exist, used mainly for hedging and topiary. Box wood is hard and is
prized for turnery.

SIMILAR TREE
Balearic Box *B. balearica* (7m) Similar to Box, but leaves paler green, less glossy and larger (to
4cm long), and twigs stiffer and thicker. Native of SW Europe, planted in Britain occasionally.

Spindle flowers and foliage

Spindle fruits

Spindle

Box flowers and foliage

Box fruits and foliage

Box trunks

Box topiary

Buckthorn *Rhamnus cathartica* (Rhamnaceae) 10m
A spreading, sometimes rather untidy deciduous tree.
BARK Dark orange-brown, becoming almost black
in older trees, but still revealing orange patches
between the numerous fissures. BRANCHES With
slender and slightly spiny shoots. LEAVES Opposite
and ovate or nearly rounded with a short pointed tip, up
to 6cm long and 4cm wide, finely toothed around the
margin and glossy green above with a pale underside;
conspicuous veins on upper surface converging
towards tip of leaf. In autumn leaves turn yellow.
REPRODUCTIVE PARTS Very small, fragrant
flowers with 4, or rarely 5, green petals. Male and
female flowers on separate trees, in small
stalked clusters, or sometimes singly, on
2-year-old shoots. Fruit black, shiny and berry-
like, about 8mm in diameter. STATUS AND
DISTRIBUTION Native across most of Europe,
including England, Wales and Ireland; absent from
Scotland. Buckthorn grows in open woods and copses,
especially on drier, calcareous soils. COMMENTS In the past,
extracts and infusions made from the berries were used in
traditional medicine, having both purgative and laxative
properties. Buckthorn and Alder Buckthorn are food plants
for larvae of the Brimstone butterfly.

Buckthorn leaves

Buckthorn fruits

Alder Buckthorn *Frangula alnus*
(Rhamnaceae) 5m
A small tree with a broadly spreading or
sometimes sprawling habit. BARK Smooth, grey
and vertically furrowed. BRANCHES Twigs have
numerous small fine hairs and are green at first,
becoming grey-brown later. Twigs and branches
opposite. LEAVES Opposite, up to 7cm long,
broadly ovate with entire margins and a short-
pointed tip. Up to 9 pairs of veins, curving
towards margin. Leaves glossy green above
and paler below, turning a clear lemon-yellow
in autumn, or redder if exposed to bright
sunlight. REPRODUCTIVE PARTS Flowers very
small and inconspicuous, rarely more than 3mm
across, greenish white with 5 petals; in small
axillary clusters opening in May or June,
sometimes later. Berry-like fruits up to 10mm in
diameter, ripening from pale green through yellow to
red and finally black. STATUS AND DISTRIBUTION A native
of much of Europe, apart from far north and drier parts of Mediterranean
region. In Britain it is found, as a native species, mainly in central and
S England and S Wales; it is scarce in Ireland and absent from
Scotland. Alder Buckthorn grows mainly in marshy woodlands and
on acidic soils. COMMENTS In the past, Alder Buckthorn was used to
make charcoal, and its even-burning properties made it particularly
suitable for gunpowder and fuses.

Alder Buckthorn leaf

Alder Buckthorn fruit and autumn leaf

Buckthorn unripe fruits and foliage

Buckthorn ripe fruits and foliage

Buckthorn bark

Alder Buckthorn bark

Alder Buckthorn flowers and foliage

Alder Buckthorn fruits and foliage

Small-leaved Lime *Tilia cordata* (Tiliaceae) 32m

A tall deciduous tree with a dense crown. Young trees have a neat, almost conical shape, but older trees are untidier, with burrs, sprouts and criss-crossed heavy branches. BARK Smooth and grey on young trees, but becoming darker and cracked in older trees, often breaking away in flakes. BRANCHES Ascending but becoming downcurved on older trees. Twigs smooth; brownish red above and olive below. Ovoid buds about 5mm long and dark red. LEAVES Up to 9cm long, rounded, with a pointed tip and heart-shaped base and a finely toothed margin. Upper surface dark shiny green and smooth; lower surface paler and smooth, but with tufts of darker hairs in the vein axils. Petiole smooth, up to 4cm long.

REPRODUCTIVE PARTS Flowers white or pale yellow and fragrant, 5-petalled, in clusters of up to 10 on a green bract 10cm long. They open in midsummer, and project at all angles from the foliage, making this an attractive tree when in flower. Fruit rounded and hard, about 6mm in diameter, held at random angles not just pendent.

STATUS AND DISTRIBUTION A native of Britain, Europe and W Asia, found in woodlands on base-rich soils. Once the dominant woodland tree in much of Britain, but now much reduced in range. Today, it is usually associated with oak and beech woodland. COMMENTS This tree harbours populations of aphids (with the consequent drip of honeydew) but not as many as on Lime. In the past, Small-leaved Lime was of commercial interest and large, ancient coppice stools are sometimes encountered. The species reproduces mainly by vegetative means – suckers and shoots.

Small-leaved Lime fruit

Small-leaved Lime leaf

Large-leaved Lime leaf

Large-leaved Lime *Tilia platyphyllos* (Tiliaceae) 40m

Tall and often narrow deciduous tree. Bole is normally free of suckers and shoots, distinguishing this species from Lime. BARK Dark grey with fine fissures in older trees, which can sometimes be ridged. BRANCHES Mostly ascending but with slightly pendent tips. Twigs reddish green and sometimes slightly downy at tip; ovoid buds, to 6mm long, dark red and sometimes slightly downy. LEAVES To 9cm long, sometimes to 15cm long, broadly ovate, with a short tapering point and irregularly heart-shaped base. Margins sharply toothed; upper surface soft and dark green, lower surface paler and sometimes slightly hairy. REPRODUCTIVE PARTS Yellowish-white flowers in clusters of up to 6 on whitish-green, slightly downy bracts, usually opening in June. Hard, woody fruit up to 1.8cm long, almost rounded or slightly pear-shaped with 3–5 ridges; a few remaining on lower branches in winter. Fruit clusters are pendent.

STATUS AND DISTRIBUTION A native of lime-rich soils in Europe; in Britain it is native to central and S England and Wales, having been introduced elsewhere; it is often planted as a street tree.

Large-leaved Lime fruit

Small-leaved Lime bark

Small-leaved Lime fruits and foliage

Small-leaved Lime

Large-leaved Lime bark

Large-leaved Lime foliage

Large-leaved Lime fruits and foliage

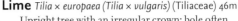

Lime bark

Crimean
Lime
leaf

Lime leaf

Lime *Tilia* × *europaea (Tilia* × *vulgaris)* (Tiliaceae) 46m

Upright tree with an irregular crown; bole often has burrs and masses of sprouts. BARK Grey-brown and ridged. BRANCHES Ascending, arching on older trees; young twigs smooth and green. Buds ovoid, 7mm long, reddish brown. LEAVES To 10cm long, broadly ovate with a short pointed tip, heart-shaped base and toothed margin; dull green above and paler below with tufts of white hairs in vein axils. REPRODUCTIVE PARTS Yellowish-white, 5-petalled, fragrant flowers, in clusters of up to 10 on a greenish-yellow bract. Fruit hard, thick-shelled and rounded; usually sterile. STATUS AND DISTRIBUTION Hybrid between Small-leaved and Large-leaved Limes. Planted and very common in towns and parks. COMMENT Suffers heavy aphid infestation, causing honeydew to rain down, so not really suitable for street planting.

Lime fruits

SIMILAR TREE

Crimean Lime *Tilia* × *euclora* (20m) Similar to Lime, but leaves dark, shiny green above with reddish hairs in vein axils below. Flower clusters with 3–7 flowers, fruits elliptical and pointed. A hybrid between Small-leaved Lime and *T. dasystyla*, planted because aphid numbers are low.

Silver-lime *Tilia tomentosa* (Tiliaceae) 30m

Broadly domed tree. BARK Grey, ridged. BRANCHES Mostly straight and ascending. Young twigs whitish and woolly, darkening with age; buds greenish brown, to 8mm long. LEAVES To 12cm long, rounded, with heart-shaped base, tapering tip and toothed margins; dark green, hairless and wrinkled above, white and downy with stellate hairs below. REPRODUCTIVE PARTS 5–10 off-white, strongly scented flowers supported by yellowish bract. Fruit to 1.2cm long, ovoid, warty and downy. STATUS AND DISTRIBUTION Native from Balkans eastwards; planted in Britain and thrives in towns. COMMENT Woolly leaves ensure no aphids, and hence no honeydew.

Silver-lime bark

Pendent
Silver-lime
leaf

SIMILAR TREE

Pendent Silver-lime *T.* 'Petiolaris' (30m) Similar to Silver-lime, but branches have pendulous tips. Leaf underside and long petiole very white and downy.

American Lime (Basswood) *Tilia americana* (Tiliaceae) 25m

Broadly columnar tree. BARK Greyish brown. BRANCHES Mostly ascending. LEAVES To 20cm long and 15cm wide, toothed margins; deep green above, paler and more glossy below with brown hair tufts in vein axils.

REPRODUCTIVE PARTS Pale yellow, 5-petalled flowers, to 1.5cm across, in clusters of up to 10 from a 10cm-long bract. Fruits hard and round. STATUS AND DISTRIBUTION Native of North America, planted in Britain occasionally.

American
Lime leaf

White
Basswood
leaf

SIMILAR TREE

White Basswood *T. heterophylla* (30m) Downy leaf undersides with buff (not brown) axillary hair tufts. Native of E USA, planted in Britain occasionally.

Lime foliage

Crimean Lime
fruits and foliage

Silver-lime
foliage

Pendent
Silver-lime foliage

Lime

Silver-lime

American Lime
foliage

White Basswood
foliage

Escallonia macrantha – planted
for hedging in Isles of Scilly

Kohuhu *Pittosporum tenuifolium* (Pittosporaceae) 10m

Stout-boled tree. BARK Smooth and dark grey. BRANCHES Densely
packed; shoots purplish black. LEAVES Oblong or elliptical, to 6cm
long and 2cm across with a wavy margin; glossy above, less shiny
below. REPRODUCTIVE PARTS Scented tubular flowers, to 1cm long,
with 5 deep purplish lobes and yellow anthers; in clusters or solitary,
in leaf axils. Fruit a rounded capsule, about 1cm long, ripening
from green to black. STATUS AND DISTRIBUTION Native of New
Zealand; planted in the British Isles but intolerant of harsh winter
conditions, thriving best in Isles of Scilly and W Cornwall.

Karo *Pittosporum crassifolium* (Pittosporaceae) 10m

Small evergreen tree or large shrub. BARK Blackish. BRANCHES
Congested. LEAVES Leathery, to 8cm long and 3cm wide, ovate to
lanceolate and blunt-tipped; dark green above, paler and woolly
below with slightly inrolled margin. REPRODUCTIVE PARTS Flowers
in lax clusters, with 5 deep red petals and yellow anthers. Fruit an
ovoid capsule, to 3cm long, matt and light green, with shiny
seeds. STATUS AND DISTRIBUTION Native of New Zealand,
tolerant of salt spray so planted for coastal hedging and
naturalised in parts of SW England.

Karo
fruit

Sea-buckthorn
leaf

French Tamarisk

Tamarix gallica (Tamaricaceae) 8m

Straggly, windswept tree. BARK Purplish brown.
BRANCHES Numerous and fine. LEAVES Greenish
blue, scale-like, to 2mm long, clasping young shoots.
REPRODUCTIVE PARTS Minute pink, 5-petalled flowers,
in tapering racemes, to 2.5cm long; each petal less than 2mm
long. Seeds wind-dispersed. STATUS AND DISTRIBUTION
Native to SW Europe; long established in the British Isles,
planted as a windbreak, or for soil stabilisation.

Sea-buckthorn

Hippophae rhamnoides (Elaeagnaceae) 11m

Multi-stemmed shrub or suckering small tree.
BARK Fissured, peeling; thorny twigs covered
with silvery scales that rub off. LEAVES To 6cm long
and 1cm wide, with silvery scales. REPRODUCTIVE PARTS
Flowers, to 3mm across, without petals, opening in March
or April. Male and female flowers on different trees. Fruits
are bright orange berries, up to 8mm long. STATUS AND
DISTRIBUTION Native of Europe, including coastal E England; planted elsewhere to
stabilise dunes, also inland for ornament. COMMENT Berries are eaten by migrant birds.

French
Tamarisk
foliage

Oleaster *Elaeagnus angustifolia* (Elaeagnaceae) 13m

Deciduous shrub or small tree. BARK Grey. BRANCHES Spiny, with silvery young twigs.
LEAVES Lanceolate, to 8cm long and 2.5cm wide; silvery below, dull green above.
REPRODUCTIVE PARTS Flowers bell-shaped, to 1cm long, singly or in pairs in leaf axils
on 2–3mm pedicels; silvery outside, yellow inside. Fruits oval, to 1cm long, yellow with
silvery scales. STATUS AND DISTRIBUTION Native of W Asia, sometimes planted in
British gardens.

Kohuhu flower and foliage

Kohuhu shrub

Karo, hedgerow shrub

Karo fruit and foliage

Karo flowers and foliage

French Tamarisk shrub

French Tamarisk flowers

Sea-buckthorn fruits and foliage

Oleaster flowers and foliage

GUMS OR EUCALYPTS *EUCALYPTUS* (FAMILY MYRTACEAE)

About 500 species occur in Australasia (none in New Zealand). Many are large, aromatic, evergreen trees that grow throughout the year. Flowers are enclosed in capsules with a cap that is eventually shed.

Cider
Gum
leaf

Cider Gum *Eucalyptus gunnii* (Myrtaceae) 30m

Medium-sized gum. BARK Smooth, readily peeling, greenish white or tinged pink. BRANCHES Upright to level. LEAVES Juvenile leaves ovate, to 4cm long, opposite with heart-shaped bases. Adult leaves ovate to lanceolate, to 7cm long, with veins sometimes prominent. REPRODUCTIVE PARTS White flowers in small clusters of 3 on slightly flattened stalks up to 8mm long. Buds cylindrical, about 8mm long with a rounded cap. Fruit, to 1cm long, is bell-shaped; slightly concave disc and up to 5 valves. STATUS AND DISTRIBUTION Native of Tasmania. Hardy in the British Isles, and much planted.

Ribbon
Gum
leaf

Broad-leaved
Kindling
Bark leaf

Southern Blue-gum *Eucalyptus globulus* (Myrtaceae) 45m

Vigorous evergreen. BARK Grey-brown, peeling. BRANCHES Mainly upright. LEAVES Juvenile leaves blue-green, opposite, stem-clasping, to 16cm long. Adult leaves dark green, alternate, to 30cm long, pendent, sometimes sickle-shaped. REPRODUCTIVE PARTS White, sessile flowers, to 4cm across; solitary or in 2s or 3s. Flower bud with pointed, warty cap. Fruit flattened, warty and ribbed. STATUS AND DISTRIBUTION Native of Tasmania, planted in Ireland for timber, elsewhere for ornament.

Ribbon Gum *Eucalyptus viminalis* (Myrtaceae) 50m

Large tree. BARK Rough, peels in long ribbons revealing smoother, pale patches. BRANCHES Mainly upright. LEAVES Juvenile leaves opposite, oblong, to 10cm long. Adult leaves alternate, to 18cm long and tapering. REPRODUCTIVE PARTS White flowers usually in clusters of 3; buds with scarlet domed caps. Fruits rounded. STATUS AND DISTRIBUTION Native of S and E Australia, grown in the British Isles for timber and ornament.

SIMILAR TREES

Broad-leaved Kindling Bark *E. dalrympleana* (40m) Broadly domed with orange and buff bark, peeling in strips. Juvenile leaves heart-shaped, adult leaves long and lanceolate. Flower buds green. Widely planted.

Snow
Gum
leaf

Snow Gum *E. pauciflora* ssp. *niphophila* (15m) Bark grey-green, peeling to reveal whiter patches. Fruits short-stalked and cup-shaped. Often planted in British gardens.

Small-leaved Gum *E. parviflora* (20m) Juvenile leaves tiny, adult leaves small and willow-like.

Urn-fruited Gum *E. urnigera* (35m) Juvenile leaves round, adult leaves ovate; planted for timber in Ireland.

Johnston's Gum *E. johnstonii* (40m) Juvenile leaves round, adult leaves narrow; planted for timber in Ireland.

Small-leaved
Gum leaf

White Peppermint-gum *E. pulchella* (15m) Leaves always narrow; planted for ornament, and self-sown on Isles of Scilly.

Small-leaved Gum fruit

White Peppermint-gum leaf

Cider Gum juvenile leaves

Cider Gum mature flowers and leaves

Ribbon Gum fruits and foliage

Cider Gum

Broad-leaved Kindling Bark foliage

Southern Blue-gum trunk

Ribbon Gum bark

Broad-leaved Kindling Bark bark

Small-leaved Gum bark

Snow Gum trunk

Small-leaved Gum fruits and foliage

Nymans
Eucryphia
leaf

Nymans Eucryphia *Eucryphia × nymansensis* (Eucryphiaceae) 17m

Narrow, columnar evergreen tree. BARK Smooth and grey. BRANCHES
Dense. LEAVES Compound, with 3 toothed, glossy dark green leaflets,
paler below, to 6cm long. REPRODUCTIVE PARTS 4-petalled white
flowers, to 7.5cm across, containing many pink-tipped stamens; in
leaf axils, opening in late summer. Fruit a small, woody capsule.
STATUS AND DISTRIBUTION Hybrid between *E. cordifolia* and
E. glutinosa, raised in Nymans Garden, Sussex; the most
frequently seen Eucryphia. *E. glutinosa*, from Chile, is
deciduous with fine autumn colours. *E. cordifolia*, known as
Ulmo, also from Chile, is evergreen; its fragrant white flowers turn
pink, then orange.

Dove Tree (Handkerchief Tree)

Davidia involucrata (Cornaceae) 20m

Slender, conical deciduous tree; stout, tapering bole. BARK Orange-brown,
peeling vertically. BRANCHES Thick; shoots smooth and brown, buds red.
LEAVES To 18cm long, heart-shaped with a pointed tip and toothed margin,
5–9 pairs of veins and a pinkish or yellow-green petiole 15cm long; dark
shiny green above, paler and downy below. REPRODUCTIVE PARTS Flowers
small, petal-less, in dense clusters of many male flowers
with purple anthers and one hermaphrodite flower;
surrounded by a large pair of white bracts, one larger
than the other, to 20cm long. Rounded fruits, to
2.5cm across, green at first, ripening to purple-
brown. STATUS AND DISTRIBUTION Native of
China, popularly planted in Britain.

Dove Tree
leaf

Dove
Tree
fruit

Tupelo (Black Gum)

Nyssa sylvatica (Nyssaceae) 25m

Broadly columnar deciduous tree. BARK
Dark grey, ridged with squarish plates.
BRANCHES Mostly level. LEAVES Ovate, to
15cm long and 8cm wide, tapering towards base.
REPRODUCTIVE PARTS Dioecious, to 1.5cm across
with downy stalks to 3cm long. Male flowers in dense, rounded clusters,
female flowers in clusters of up to 4 flowers. Fruit about 2cm long,
egg-shaped and bluish black. STATUS AND DISTRIBUTION Native
of E North America, grown in Britain for its exciting yellow,
orange and red autumn colours. Favours warm, sheltered areas.

Tupelo
leaf

Chilean Myrtle *Luma apiculata* (Myrtaceae) 12m

Broadly spreading evergreen tree. BARK Bright orange, flaking
to reveal white patches beneath. BRANCHES Slightly spreading,
also with peeling bark low down. LEAVES Pointed, elliptical,
to 2.5cm long, with entire margins; glossy dark green above,
paler below. Aromatic. REPRODUCTIVE PARTS White, 4-petalled
flowers, to 2cm across, with many stamens bearing yellow
anthers; solitary, in leaf axils, opening in late summer. Fruits
fleshy, berry-like, to 1cm long, ripening from red to black.
STATUS AND DISTRIBUTION Native of S Andes, grown in the
British Isles for ornament; naturalised in south-west of region.

Nymans Eucryphia foliage

Nymans Eucryphia flowers

Dove Tree flowers and foliage

Dove Tree flower

Tupelo autumn colours

Chilean Myrtle bark

Chilean Myrtle foliage

Dogwood *Cornus sanguinea* (Cornaceae) 4m

Dogwood leaf

Shrub, or sometimes a small tree on a slender bole. BARK Grey and smooth. BRANCHES Dark red winter twigs distinctive after leaves have fallen. LEAVES Opposite, oval and pointed, with entire margins and 3–4 pairs of prominent veins. If a leaf is snapped and 2 halves gently pulled apart, stringy latex appears where veins were broken and connects 2 halves of leaf. Leaves reddish green above, becoming a rich, deep red in autumn. REPRODUCTIVE PARTS White flowers small, but grouped in large terminal clusters. Fruit a blackish, rounded berry borne in clusters. STATUS AND DISTRIBUTION Widespread native across Europe; in the British Isles it is commonest in England, but planting schemes, especially along roadside verges, ensure that it occurs outside its natural range. Prefers calcareous soils, usually growing in thickets, hedgerows and woodland edges; quick to colonise new ground as birds carry the seeds.

SIMILAR TREES

LEFT: Red-osier Dogwood twig
RIGHT: Red-osier Dogwood leaf
BELOW: Cornelian-cherry leaf

Red-osier Dogwood *C. sericea* (3m) Shrub with bright red, or yellow, twigs in winter. Leaves ovate, to 10cm long, with pointed tips. Native of North America, planted in Britain, sometimes naturalised.

White Dogwood *C. alba* (3m) Twigs bright red in winter. Leaves more rounded than those of *C. sericea*. Native of E Asia, planted for ornament.

Cornelian-cherry *Cornus mas* (Cornaceae) 8m

Small, spreading deciduous tree with an untidy crown. BARK Reddish brown. BRANCHES Mostly level, ending in numerous greenish-yellow, slightly downy twigs. LEAVES Opposite, short-stalked, ovate and pointed, to 10cm long and 4cm wide with rounded bases; dull green above and slightly downy with entire margins. REPRODUCTIVE PARTS Flowers in small stalked heads, about 2cm across, consisting of up to 25 small yellow flowers, each about 4mm across. Flowers open early in year, well before leaves. Fruit a short-stalked, pendulous, bright red, fleshy berry, to 2cm long, with pitted apex and acid taste. STATUS AND DISTRIBUTION Native of scrub and open woodlands in central and SE Europe, grown in Britain for its winter flowers and edible fruits. Naturalised occasionally.

Strawberry Dogwood *Cornus kousa* (Cornaceae) 15m

Strawberry Dogwood leaf

Strawberry Dogwood fruit

A columnar to pyramidal deciduous tree. BARK Reddish brown, peeling off in patches in older trees. BRANCHES Tangled. LEAVES Ovate, to 7.5cm long and 5cm across, with a tapering point and wavy margin; dark green above, smooth below with patches of brown hairs in vein axils. REPRODUCTIVE PARTS Small yellowish-white or greenish flowers clustered together in compact rounded heads, surrounded by 4 large yellowish-white or pink-tinged bracts; opening in early summer, followed by bunches of tiny, edible fruits that collectively look like strawberries. STATUS AND DISTRIBUTION Native of Japan, and a garden tree in Britain and Europe.

Dogwood bark

Dogwood flowers and foliage

Dogwood fruits and foliage

Cornelian-cherry flowers

Dogwood, fruiting shrub

Cornelian-cherry fruits and foliage

Strawberry Dogwood flower and foliage

Strawberry Dogwood fruit

Rhododendron *Rhododendron ponticum* (Ericaceae) 5m

Evergreen ornamental shrub. BARK Reddish and scaly. BRANCHES Dense and tangled. LEAVES Shiny, leathery, elliptical and dark green. REPRODUCTIVE PARTS Flowers 4–6cm long, bell-shaped and pinkish red; in clusters in May and June. Fruits dry capsules containing numerous flat seeds. STATUS AND DISTRIBUTION Native of Asia and SE Europe, widely planted in the British Isles and naturalised in some areas. Favours acid, damp soils. COMMENTS This species' invasive habits, and its ability to exclude native flora, mean that it is often controlled.

Strawberry Tree *Arbutus unedo* (Ericaceae) 9m

Small, spreading evergreen tree with a short bole and a dense, domed crown. BARK Reddish, peeling away in shreds that turn brown. BRANCHES Often ascending and twisted; twigs slightly hairy and reddish. LEAVES To 11cm long, with a prominent midrib, and margins either sharply toothed or entire. Dark glossy green above, paler below; petiole 1cm long, and usually red and hairy. REPRODUCTIVE PARTS Flowers in pendulous clusters, late in year at same time as fruits from previous year; flowers white, to 9mm long, sometimes tinged pink or green. Fruit a round berry, to 2cm across; warty skin ripening from yellow through orange to deep red; flesh acidic. STATUS AND DISTRIBUTION Main native range is SW Europe and Mediterranean; also occurs naturally in SW Ireland in open woods and thickets. Planted widely elsewhere.

Strawberry Tree fruits

ABOVE: Rhododendron leaf

RIGHT: Strawberry Tree leaf

Strawberry Tree flowers

Date-plum *Diospyros lotus* (Ebenaceae) 14m

Small, deciduous tree. BARK Grey or pink-tinged, broken by fissures into small plates. BRANCHES Spreading. LEAVES Ovate or lanceolate with pointed tips and untoothed margins, dark glossy green above and greyer below; young leaves downy above. REPRODUCTIVE PARTS Flowers bell-shaped and salmon-pink or orange-yellow. Male and female flowers on separate trees, males clustered and smaller than single females, which are about 5mm long. Fruit an edible berry 2cm long; ripening from green, through yellow-brown to blue-black. STATUS AND DISTRIBUTION Native of SW Asia, planted in Britain occasionally.

Date-plum bark

Date-plum leaf

Snowbell Tree *Styrax japonica* (Styracaceae) 12m

Spreading deciduous tree or large shrub. BARK Smooth, dark greyish brown, fissured with age. BRANCHES Mostly ascending with zigzag slender shoots and purple-tinged buds. LEAVES Elliptical to ovate, to 10cm long with narrow bases, pointed tips, and finely toothed margins; deep glossy green above, turning yellow or red in autumn. REPRODUCTIVE PARTS White, lightly scented, 5-petalled flowers, to 1.5cm long, hanging singly or in small clusters from branches; open in midsummer. Fruit an egg-shaped berry, to 1.5cm long. STATUS AND DISTRIBUTION Native of E Asia, planted occasionally for ornament in the British Isles.

Snowbell Tree fruit

Rhododendron flowers and foliage

Strawberry Tree flowers and foliage

Date-plum fruits and foliage

Strawberry Tree flowers, fruits and foliage

Snowbell Tree flowers and foliage

Snowbell Tree fruits and foliage

Date-plum ripe fruits

ASHES *FRAXINUS* (FAMILY OLEACEAE)

Members of this genus all have opposite pinnate leaves. Most species have small, wind-pollinated flowers, but a few have scented flowers with petals.

Ash *Fraxinus excelsior* (Oleaceae) 40m

A large deciduous tree with a straight bole and a high, open, domed crown. BARK Smooth and pale grey in young trees, but becoming vertically fissured in older trees, although remaining grey; the true colour is often obscured by large colonies of lichens, which grow well on mature Ash boles in unpolluted areas. BRANCHES Mostly ascending, terminating in grey twigs that are flattened at the nodes and are tipped with conical sooty black buds. LEAVES Pinnate; up to 35cm long with a flattened central rachis, which may be hairy, bearing 7–13 ovate-lanceolate, pointed and toothed leaflets, each up to 12cm long. Upper surface usually dark green; lower surface paler with densely hairy midribs. The leaves are among the latest on any native tree to open in the spring. They turn pale yellow-green for a short time in autumn before falling quickly. REPRODUCTIVE PARTS Very small, purple flowers in clusters near tips of twigs in spring. Male and female flowers separate, and mostly on separate trees, but some trees have both on separate branches. Fruits single-winged 'keys', hanging in bunches, starting off green and ripening to brown, and usually persisting until after the leaves have fallen. STATUS AND DISTRIBUTION A widespread native of most of Europe, including Britain and Ireland. It prefers calcareous or base-rich soils and grows well on limestone uplands, but also thrives on heavy base-rich clays, near the sea and in cities. Woodland trees have the best shape, while hedgerow trees tend to be stunted and misshapen. The species is widely planted for timber and occasionally for shelter-belts. COMMENTS Ash is found in many ancient semi-natural woodlands, often having been coppiced over the centuries with the result that huge stools, or rings of stools, are formed. In many locations it is these, rather than the oaks alongside which they often grow, that are the oldest trees – some stool rings can have a diameter of 4m or more and exceed 800 years in age. Ash woodlands characteristically have a light, airy canopy, which means that a rich ground flora can often flourish and carpets of Bluebells, Wood Anemones and Dog's Mercury are not uncommon. Ash timber is white, durable and easily worked, and is popular for making farm implements and furniture. Its inherent strength makes it ideal for the handles of pickaxes, spades and hammers, and it also makes excellent firewood, particularly since it does not need as much seasoning as other timbers.

Ash leaf

Ash winged seed

Ash seed cluster

Ash bark

Ash foliage

Ash flowers

Ash

Ash fruits and foliage

Ancient coppiced Ash stool

Narrow-leaved Ash *Fraxinus angustifolia* (Oleaceae) 25m

Narrow-leaved
Ash leaf

Resembles Ash but crown is tall and untidy. BARK Fissured; warty in older trees. BRANCHES Sparse and ascending. Buds dark brown and hairy. LEAVES Much narrower, toothed and long-pointed leaflets than Ash, especially in older trees. REPRODUCTIVE PARTS Hermaphrodite flowers open before leaves. Winged fruits hang in small clusters. STATUS AND DISTRIBUTION Native of S Europe, planted in Britain for ornament.

Caucasian
Ash leaf

SIMILAR TREE
Caucasian Ash *F. oxycarpa* (25m) Bark silvery grey. Leaves glossy with white hairs on base of midribs beneath; in var. 'Raywood' rich purple in autumn. Native of SE Europe and Asia Minor, planted occasionally.

Manna Ash *Fraxinus ornus* (Oleaceae) 24m

Medium-sized deciduous tree with a flattish crown. BARK Smooth dark grey, sometimes almost black. BRANCHES With smooth, grey twigs, sometimes tinged yellow, ending in greyish, white-bloomed buds. LEAVES Opposite, pinnate, to 30cm long with up to 9 ovate, toothed leaflets, each to 10cm long and downy, with white or brown hairs on veins beneath. REPRODUCTIVE PARTS Showy, creamy white, fragrant flowers opening with leaves and each with 4 petals about 6mm long; hanging in clusters about 20cm across. Narrow-winged fruits, to 2cm long, hanging in dense clusters. STATUS AND DISTRIBUTION Native of central and S Europe and SW Asia, and planted in the British Isles as a street tree or for ornamental value.

Caucasian
Ash fruit

Manna
Ash leaf

White Ash *Fraxinus americana* (Oleaceae) 30m

Broadly columnar deciduous tree. BARK Grey-brown, intricately ridged. BRANCHES Upright, with straight shoots. LEAVES Leaflets smooth above and white beneath; blades do not continue down the petiole. Autumn colour unreliable (in NW Europe) but can be impressive, with purple-bronze leaves. REPRODUCTIVE PARTS Similar to Ash. STATUS AND DISTRIBUTION Native to E North America, planted in the British Isles occasionally.

Manna
Ash fruit

Red Ash *Fraxinus pennsylvanica* (Oleaceae) 25m

White
Ash leaf

Similar to Ash but generally smaller. BARK Deeply furrowed, reddish brown. BRANCHES Stout with hairy twigs and brown winter buds. LEAVES Opposite, pinnate, to 22cm long with 7, rarely 9, oval, pointed leaflets. Each leaflet to 15cm long, toothed and pointed, 2 sides of blade not matching on petiole; undersides usually hairy. REPRODUCTIVE PARTS Flowers opening before leaves in hairy clusters in leaf axils; sexes usually on separate trees. Male flowers red, female flowers greenish. Both sexes lacking petals, but female flowers with 4 sepals. Single-winged fruits up to 6cm long. STATUS AND DISTRIBUTION Native of E North America; planted occasionally in Britain and Ireland.

Red Ash
leaf

Red Ash
fruit

Caucasian Ash bark

Narrow-leaved Ash foliage

Manna Ash flowers and foliage

Manna Ash autumn colours

White Ash foliage

Red Ash foliage

Red Ash flowers and foliage

Wild Privet
leaf

Wild Privet
fruit

Wild Privet
fruit

Wild Privet *Ligustrum vulgare* (Oleaceae) 5m
A much-branched, semi-evergreen spreading
shrub, very rarely reaching the size of a small
tree. BARK Reddish brown with vertical scar-
like gashes. BRANCHES Dense and much
divided, with downy young twigs. LEAVES Shiny,
untoothed, oval and opposite. REPRODUCTIVE PARTS
Flowers 4–5mm across, creamy white, fragrant,
4-petalled; in terminal spikes and appearing in
May and June. Fruits shiny, globular and
poisonous, ripening black in the autumn; in
clusters. STATUS AND DISTRIBUTION Found wild in
limestone and chalk areas in much of Europe, where it can
form dense scrub. Locally common as a native species in S
and central England and Wales, but generally scarce
elsewhere in the British Isles. However, it is also widely
planted and naturalised and this obscures its natural range;
it is found in hedgerows and areas of scrub. COMMENTS Wild
Privet is often used as a hedging plant. When clipped to form
a neat hedge, it is seldom able to form flowers and fruits. However,
in the wild the latter are usually produced in good numbers and are
eagerly consumed by birds, such as thrushes, which in turn
spread the seeds. The leaves are the food plant for larvae of the
Privet Hawkmoth (*Sphinx ligustri*), as are those of Garden Privet
and Lilac.

SIMILAR TREE
Garden Privet *L. ovalifolium* (3m) Similar but with
leaves rounded-oval and more evergreen. A native
of Japan, often planted in gardens for hedging,
sometimes in one of its variegated forms.

Garden Privet
leaf

Lilac *Syringa vulgaris* (Oleaceae) 7m
A small deciduous tree, but sometimes little more
than a multi-stemmed shrub with a rounded crown
and a short bole surrounded by suckers. BARK Greyish
and spirally fissured in older trees. BRANCHES Usually a
mass of ascending branches. Twigs rounded and shiny
greenish brown. LEAVES Opposite, up to 10cm long and with a short
petiole; ovate or slightly heart-shaped with entire margins and a
slightly leathery feel. Usually yellowish green with a smooth
surface. REPRODUCTIVE PARTS Fragrant lilac flowers in dense,
paired conical spikes, up to 20cm long, arising from the apical
leaf axils; at their best in May and June. Individual flowers up to
1.2cm long and 4-lobed. Fruit is a pointed, ovoid capsule up to
1cm long. STATUS AND DISTRIBUTION A native of rocky
hillsides in the Balkans, growing in open thickets and scrub, but
long cultivated in the rest of Europe for its attractive fragrant
flowers. In Britain and Ireland, it is a popular garden plant and
frequently naturalised as well, spreading by vegetative means (mainly
suckers) rather than seed.

Lilac leaf

Wild Privet bark

Wild Privet flowers and foliage

Wild Privet fruits

Wild Privet

Privet Hawkmoth caterpillar on Garden Privet

Lilac flowers and foliage

Indian Bean Tree (Southern Catalpa)

Catalpa bignonioides (Bignoniaceae) 20m

Medium-sized deciduous tree with a short bole. BARK Greyish brown and scaly. BRANCHES Mostly spreading; smooth, stout twigs tipped with very small orange-brown buds. LEAVES Long-stalked, large and broadly ovate, to 25cm long and 20cm across, with heart-shaped bases and short-pointed tips; margins untoothed, upper surface smooth and lower surface downy. Leaves tinged with purple and downy when young, becoming a lighter, almost transparent green when mature. Usually late to open and early to fall. REPRODUCTIVE PARTS Flowers 5cm long, and an open bell shape with 2 lips; petals white with purple and yellow spots; in large showy panicles in midsummer. Fruit a long, slender bean-like pod, to 40cm long, hanging from branches long after leaves have fallen; contains many inedible flat, papery seeds, to 2.5cm long. STATUS AND DISTRIBUTION Native of SE USA, planted and quite common in many large cities in Britain, including London.

Indian Bean Tree leaf

Indian Bean Tree fruit

Yellow Catalpa

Catalpa ovata (Bignoniaceae) 12m

Very similar to Indian Bean Tree, best distinguished by comparing leaves. BARK Grey-brown and scaly. BRANCHES Spreading. LEAVES Dark green and often pentagonal with a short point on each corner; large, to 25cm in each direction, with a heart-shaped base. REPRODUCTIVE PARTS Off-white flowers, to 2.5cm across, tinged yellow with red spots inside; in spikes about 25cm long. The pod is about 25cm long. STATUS AND DISTRIBUTION A native of China, introduced to the British Isles for ornament and rarely seen outside parks and gardens.

Yellow Catalpa leaf

Foxglove Tree leaf

Foxglove Tree

Paulownia tomentosa (Scrophulariaceae) 15m

Small deciduous tree with an upright habit. BARK Smooth and grey. BRANCHES Mostly level. LEAVES Huge, to 35cm across. Very broadly ovate with heart-shaped bases and tapering tips, often with forward-pointing side lobes. Upper surface light green and hairy, lower surface grey-green and much more hairy. Petiole about 15cm long, but sometimes as much as 45cm long, and very downy. REPRODUCTIVE PARTS Flowers in lax upright spikes about 30cm long; each flower brown and downy in bud, but opening to become violet, with a yellowish tinge inside the corolla tube. Flower about 6cm long with 5 spreading lobes. Fruit a short-stalked, ovoid capsule about 5cm long with a tapering tip and glossy green outer skin; it splits open to release many small, whitish, winged seeds. STATUS AND DISTRIBUTION A native of the mountains of China, brought as an ornamental tree to Britain and Ireland, where it is often seen in large gardens and sometimes as a street tree.

Indian Bean Tree flowers and foliage

Indian Bean Tree foliage

Yellow Catalpa flowers

Yellow Catalpa

Foxglove Tree bark

Foxglove Tree fruits

Foxglove Tree foliage

Elder *Sambucus nigra* (Caprifoliaceae) 10m

A small deciduous, often rather untidy tree or a large shrub; the bole is normally short and an old bole often has fast-growing young shoots emerging from it. BARK Deeply grooved and furrowed bark, greyish brown and often taking on a corky texture in older specimens. BRANCHES Numerous, spreading and twisted. Branches and twigs have soft white pith in the centre. LEAVES Opposite and compound with 5–7 (occasionally 9) pairs of leaflets, each one up to 12cm long, ovate and pointed with a sharply toothed margin and slightly hairy underside. Green through the summer but sometimes turning a deep plum-red before falling in autumn. Crushed leaves have an unpleasant 'catty' smell. REPRODUCTIVE PARTS Sickly sweet scented flowers in dense, flat-topped clusters up to 24cm across; individual flowers small and composed of 3–5 white petals and anthers. Fruit a rounded, shiny black berry, often produced in great numbers in pendulous heads. STATUS AND DISTRIBUTION A widespread and common native of woodlands, hedgerows and scrub over much of Europe except the extreme north. In Britain and Ireland, it is widespread and common on waste ground and wherever the soil has a high nitrogen content. COMMENTS Treated as a weed and grubbed out in some areas, it is highly prized and even cultivated in others for its edible flowers and fruits. Elderflowers are popular with nectar-feeding insects. Elderberries are eagerly consumed by birds and the seeds are readily dispersed. Being resistant to rabbit nibbling, and benefiting from their droppings, Elder often occurs near warrens. Elder wood, from old and mature trunks, was used in the past for whittling and carving. In winter, old stems are often adorned by a peculiarly ear-like fungus called Jelly Ear *Auricularia auricula-judae*.

Elder leaf

Elder var. *laciniata* leaf

Red-berried Elder leaf

SIMILAR TREE
Red-berried Elder *S. racemosa* (3m) Similar to Elder but with red berries. Planted in gardens and naturalised occasionally.

Dwarf Elder *Sambucus ebulus* (Caprifoliaceae) 2m

Unpleasant-smelling deciduous shrub or very occasionally a small tree. BARK Reddish brown. BRANCHES Arching and grooved, turning red in September. LEAVES Compound, divided into 7–13 narrow leaflets, narrower than those of Elder. REPRODUCTIVE PARTS Flowers 3–5mm across and pinkish white; in flat-topped clusters, 8–15cm across and appearing from June to August. Fruits black, poisonous berries in clusters. STATUS AND DISTRIBUTION Doubtfully native to Britain and Ireland; patchily distributed and generally scarce, being found mainly in the south of the region. Grows in hedgerows and scrub, and on roadside verges. COMMENTS In the past, Dwarf Elder was used to make a black dye, which may have been why it was introduced (if indeed it was) to Britain.

Dwarf Elder leaf

Elder flowers

Elder

Elder bark

Elder bark

Elder fruits

Jelly Ear on Elder branch

Dwarf Elder fruits

Dwarf Elder flowers

Guelder-rose leaf

Guelder-rose *Viburnum opulus* (Caprifoliaceae) 4m

Small, sometimes rather spreading deciduous tree. BARK Reddish brown. BRANCHES Sinuous, if growing in a crowded situation in woodland; twigs smooth, angular and greyish. LEAVES Opposite, to 8cm long, with 3–5 irregularly toothed lobes and thread-like stipules. They often turn a deep wine-red in the autumn. REPRODUCTIVE PARTS White flowers in flat heads, resembling a lacecap hydrangea: large, showy but sterile outer flowers, with smaller, fertile flowers in centre. Fruit a rounded, glistening, translucent red berry, hanging in clusters on tree after leaves have fallen. STATUS AND DISTRIBUTION Native of damp woodlands, hedgerows and thickets across much of Europe. Widespread in Britain and Ireland, favouring calcareous and neutral soils.

Wayfaring-tree leaf

Wayfaring-tree *Viburnum lantana* (Caprifoliaceae) 6m

Small, spreading deciduous tree. BARK Brown. BRANCHES With rounded, greyish, hairy twigs; through a hand-lens hairs look star-shaped. LEAVES Opposite, to 14cm long and ovate, rough to touch and with toothed margins; undersides thickly hairy with more stellate hairs. REPRODUCTIVE PARTS Flowers in midsummer in rounded flower heads about 10cm across comprising many small white flowers, all of which are fertile; each flower about 8mm across with 5 white petals. Fruits flattened oval berries about 8mm long, red at first but ripening to black; ripening is staggered on the same fruiting head, giving a striking mixture of red and black berries side by side. STATUS AND DISTRIBUTION Native to most of Europe except for the extreme north, growing mainly on drier chalky soils at the edges of woods and in hedgerows and scrub patches. In Britain, native to S England and S Wales and occasionally planted elsewhere. Larvae of the Orange-tailed Clearwing *Synanthedon andrenaeformis* (a moth) live in the stems.

Laurustinus leaf

Laurustinus *Viburnum tinus* (Caprifoliaceae) 7m

Small evergreen tree with attractive glossy foliage and flowers produced freely in winter. BARK Brown. BRANCHES With faintly angled, slightly hairy twigs. LEAVES Opposite, to 10cm long and oval, sometimes lanceolate, with entire margins, a dark green glossy upper surface and a paler, slightly hairy lower surface. REPRODUCTIVE PARTS Pink and white flowers in branched, rounded clusters, to 9cm across; individual flowers about 8mm across with 5 petals, pink outside and white inside. Rounded fruits about 8mm long and steely blue when ripe. STATUS AND DISTRIBUTION Native of Mediterranean region, but hardy and so widely planted in Britain as a garden shrub or tree, also used for hedging and shelter; naturalised occasionally.

Guelder-rose flowers and foliage

Guelder-rose fruits

Guelder-rose fruiting shrub

Wayfaring-tree flowers and foliage

Wayfaring-tree

Wayfaring-tree fruit and foliage

Laurustinus flowers and foliage

Laurustinus

New Zealand
Holly leaf

New Zealand Holly *Olearia macrodonta* (Asteraceae) 3m

Bushy shrub. BARK Brown and stringy. BRANCHES Much divided. LEAVES Ovate with sharply toothed margins. REPRODUCTIVE PARTS Flowers whitish, daisy-like, in dense clusters. STATUS AND DISTRIBUTION Native of New Zealand. Planted in the British Isles for hedging and self-sown; not hardy so found mainly in coastal W Britain.

Akiraho
leaf

SIMILAR TREES

Akiraho *O. paniculata* (4m) Similar but with smaller leaves that have undulate margins. Native to New Zealand, planted for hedging in coastal SW England.
Ake-ake *O. traversii* (6m) Leaves oval, entire. Native to Chatham Island, grown and naturalised on Isles of Scilly.

Ake-ake
leaf

Cabbage Palm *Cordyline australis* (Agavaceae) 13m

A superficially palm-like evergreen. Trees that have flowered have a forked trunk with a crown of foliage on top of each fork. BARK Pale brownish grey, ridged and furrowed. LEAVES Dense masses of long, spear-like, parallel-veined leaves, to 90cm long and 8cm wide, crowning tall, bare trunks. Upper leaves mostly erect, lower leaves hanging down to cover top of trunk. REPRODUCTIVE PARTS Flowers in midsummer in large spikes, to 1.2m long, comprising numerous small, fragrant, creamy-white flowers, each about 1cm across, with 6 lobes and 6 stamens. Fruit a small, rounded bluish-white berry, about 6mm across, containing several black seeds. STATUS AND DISTRIBUTION Native of New Zealand, planted elsewhere for ornament. In Britain it survives quite far north, as long as there is some protection from severe cold, and tolerates a range of soil types. Often used to create the illusion of subtropical conditions in coastal resorts.

Unrelated New Zealand
Broadleaf *Griselinia
littoralis* (right) is used
as a hedging shrub in
SW Britain

New Zealand
Broadleaf

Chusan Palm *Trachycarpus fortunei* (Arecaceae) 14m

Palm whose tall bole is covered with persistent fibrous leaf bases that hide the bole itself. LEAVES Palmate and up to 1m in diameter, split almost to the base; segments stiff and pointed, usually bluish green on the underside and dark green above. Petiole to 50cm long, toothed on the margins, with the base hidden by dense brown fibres. REPRODUCTIVE PARTS Fragrant yellow flowers on a long, branched spike; males and females on different trees. Protected before opening by enveloping white or brown bracts. 6 yellow segments in the flower, the inner 3 being the largest. Large numbers of 3-lobed fruits, 2cm long and tinged purple, in late summer. STATUS AND DISTRIBUTION A native of China, introduced to Europe as an ornamental tree, and common on roadsides and in parks and gardens. One of the hardiest palms, and will survive our climate, so it is also found in many coastal resorts in the milder parts of Britain and Ireland.

leaf upperside   leaf underside

New Zealand Holly flowers and foliage

Ake-ake foliage

Ake-ake hedge

Ake-ake flowers

Akiraho foliage

Cabbage Palm

Cabbage Palm flowers and foliage

Cabbage Palm bark

Chusan Palm foliage

Chusan Palm bark

FURTHER READING

British Wildlife. A magazine for the modern naturalist (6 issues per year). British Wildlife Publishing.

BSBI Handbooks. A series of identification guides to selected plant groups or families. Botanical Society of the British Isles.

Edlin, H.L. (1970). *Trees, Woods and Man* (New Naturalist Series, no. 32). Collins.

Fitter, A. (1978). *An Atlas of the Wild Flowers of Britain and Northern Europe*. Collins.

Fitter, R., Fitter, A. and Blamey, M. (1974). *The Wild Flowers of Britain and Northern Europe*. Collins.

Gibbons, R. and Brough, P. (1992). *The Hamlyn Photographic Guide to the Wild Flowers of Britain and Northern Europe*. Hamlyn.

Johnson, O. and More, D. (2004). *Tree Guide*. HarperCollins.

Morton, A. (2004). *Tree Heritage of Britain and Ireland*. Airlife.

Preston, C.D., Pearman, D.A. and Dines, T.D. (2002). *New Atlas of the British & Irish Flora*. Oxford University Press.

Rose, F. (illustrated by Davis, R.B., Mason, L., Barber, N., and Derrick, J.) (1981). *The Wild Flower Key*. Warne.

Spooner, B. and Roberts, P. (2005). *Fungi* (New Naturalist Series, no. 96). HarperCollins.

Stace, C. (1997). *New Flora of the British Isles* (second edition). Cambridge University Press.

Tubbs, C.R. (1986). *The New Forest* (New Naturalist Series, no. 73). Collins.

USEFUL ORGANISATIONS

The following list includes organisations that I have found useful in the context of trees, woodland and conservation:

Ancient Tree Forum www.woodland-trust.org.uk/ancient-tree-forum
Botanical Society of the British Isles (BSBI) www.bsbi.org.uk
Bramley Frith www.bramleyfrith.co.uk
The Forestry Commission www.forestry.gov.uk
The National Trust www.nationaltrust.org.uk
The National Trust Heritage Trees www.nationaltrust.org.uk/main/w-chl/
    w-countryside_environment/w-woodland/w-woodland-heritage_trees.htm
The National Trust for Scotland www.nts.org.uk
Royal Society for the Protection of Birds (RSPB) www.rspb.org.uk
The Wildlife Trusts www.wildlifetrusts.org.uk
The Woodland Trust www.woodland-trust.org.uk

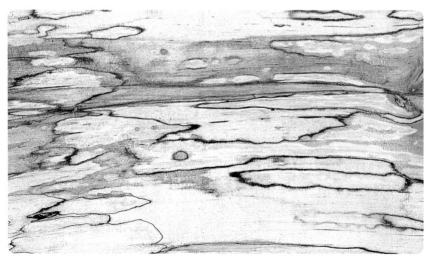

The beautiful patterns in spalted birch timber are created by fungal decay.

INDEX

PICTURE CREDITS